The Biology of Beauty

The Biology of Beauty
The Science behind Human Attractiveness

RACHELLE M. SMITH

An Imprint of ABC-CLIO, LLC
Santa Barbara, California • Denver, Colorado

Copyright © 2018 by ABC-CLIO, LLC

All rights reserved. No part of this publication may be reproduced, stored in a retrieval system, or transmitted, in any form or by any means, electronic, mechanical, photocopying, recording, or otherwise, except for the inclusion of brief quotations in a review, without prior permission in writing from the publisher.

Library of Congress Cataloging-in-Publication Data

Names: Smith, Rachelle M., author.
Title: The biology of beauty : the science behind human attractiveness / Rachelle M. Smith.
Description: Santa Barbara : Greenwood, 2018. | Includes bibliographical references and index.
Identifiers: LCCN 2018001414 (print) | LCCN 2018015682 (ebook) | ISBN 9781440849893 (ebook) | ISBN 9781440849886 (alk. paper)
Subjects: LCSH: Aesthetics—Physiological aspects. | Beauty, Personal. | Interpersonal attraction.
Classification: LCC BH301.P45 (ebook) | LCC BH301.P45 S65 2018 (print) | DDC 153.7/58—dc23
LC record available at https://lccn.loc.gov/2018001414

ISBN: 978-1-4408-4988-6 (print)
       978-1-4408-4989-3 (ebook)

22 21 20 19 18   1 2 3 4 5

This book is also available as an eBook.

Greenwood
An Imprint of ABC-CLIO, LLC

ABC-CLIO, LLC
130 Cremona Drive, P.O. Box 1911
Santa Barbara, California 93116-1911
www.abc-clio.com

This book is printed on acid-free paper ∞

Manufactured in the United States of America

Contents

Preface	vii
Part I: Understanding Beauty	**1**
1. Defining Beauty	3
2. The Benefits of Beauty	31
3. Buying Beauty	53
4. Changes in Beauty Trends over Time	81
5. Evolution's Impact on Modern Attraction: The Interaction of Genes and Environment	97
6. The Impact of Attractiveness on Behavior and Relationship Satisfaction	117
7. Psychological Effects of the Preoccupation with Beauty	141
Part II: Beauty from Head to Toe	**165**
Head, Facial, and Body Hair	165
Skin	176
Face Shape and Structure	183
Eyes	189

Nose	194
Ears	195
Lips	197
Teeth	200
Hands	203
Body Shape and Proportions	204
Breasts and Buttocks	215
Muscularity	217
Inguinal Crease	218
Male Genitalia	218
Feet	219
Hormonal Influences on Attraction	221
Psychological Traits	231
Glossary	233
References and Further Reading	241
Index	253

Preface

From the time of the ancient Egyptians and Greeks, human attractiveness has been analyzed, sought, and revered. Specific proportions such as the golden ratio were initially thought to underlie perfection. Even then, researchers believed there was a fundamental relationship between attractiveness and other positive health, behavioral, and personal qualities. Current scientific research continues to explore what it is that makes someone physically attractive and how those traits relate to other qualities. Many psychology and evolutionary researchers have built careers studying beauty and physical attraction around the world, examining the underpinnings of what is beautiful and how beauty relates to success, physical and mental health, opportunities, friendships, sexual relationships, behavior, and marriage. Their research has added to a foundation of understanding what it means to be attractive and the biological underpinnings of beauty. Research clearly demonstrates a universal conception of beauty that is largely consistent throughout the world regardless of age, culture, political system, religion, mating style, or government structure.

The goal of this book is to explore these similarities to discover the underlying biology of beauty. The book explores basic questions surrounding what is considered beautiful, the benefits of beauty, how individuals try to make themselves more attractive (now and in the past), beauty trends in the United States, the evolutionary explanation for attraction, the impact of attractiveness on relationships, and the psychological effects of a preoccupation with attractiveness.

Part I of this book reviews empirical research that reveals the major contributors to physical attractiveness. Chapter 1 breaks down the individual aspects that constitute objective and subjective beauty. These range from traits such as symmetrical proportions of the body and face, and signs of youth and health, to interpersonal traits such as personality, ambition, and familiarity. Once beauty is defined, Chapter 2 explores the biological and social benefits of possessing such traits. Chapter 3 assesses the research on how individuals of average attractiveness can make themselves more beautiful and the impact such changes have on reaping the benefits of beauty. Chapter 4 looks at changes in standards of attractiveness throughout time and the media's impact on beauty trends in the Unites States. Chapter 5 examines the role of evolution on attraction and relationships. Chapter 6 assesses how attractiveness impacts relationship satisfaction and longevity, and Chapter 7 analyzes the psychological effects that result from the quest to meet beauty standards.

In Part II, specific body parts are examined to assess the research on what makes each attractive. Part II also includes cultural sidebars that examine specific and conflicting beauty trends in cultures around the world and throughout history.

Part I
Understanding Beauty

1
Defining Beauty

Physical attractiveness is a combination of many individual features. This chapter explores multiple factors that have been found to influence one's level of attractiveness. While personal attraction certainly may vary from individual to individual based on personal experiences and preferences and developmental and cultural influences, the foundation for what we, as humans, tend to find beautiful is arguably a by-product of our genetic makeup and is largely universal. One of the ways we know this is that cross-cultural ratings of photographs are highly consistent with regard to which are the most beautiful, and even babies seem to be able to identify a beautiful face. When researcher Judith H. Langlois and colleagues from the University of Texas at Austin presented babies with photos of more and less attractive people (as rated by adults), babies chose to look longer at the more attractive faces. Furthermore, viewing a beautiful face activates the reward centers in the brain in adults, regardless of the viewer's race, nationality, or culture.

Based on Dr. Langlois's research, as well as others' research on beauty, we know that there are genetic influences on what we find to be beautiful. This is demonstrated by a consistency in ratings across different societies. However, other research demonstrates that cultural influences also play a role, and perception of attractiveness is a tug of war between objective and subjective qualities. Objective qualities can and have been tirelessly studied and measured. These include physical traits such as symmetry, age, body shape and proportions, facial features, body scents, and vocal qualities. Biologists, anthropologists, and evolutionary psychologists have found that specific proportions, features, and vocal pitches directly correlate to the level of attractiveness and to the probability of reproductive success, varying predictably by gender and culture.

Subjective qualities, however, also play a role in perceived attractiveness. Social psychologists study how personality, ambition, similarity, agreeableness, stability, and familiarity can all taint the perception of beauty and influence the viewer independently of actual physical traits. The level of attractiveness of these qualities tends to vary predictably depending on the values of the culture in question, and these values substantially impact the ratings of overall attractiveness. Physical attractiveness, therefore, is not entirely physical.

WHY ARE THERE UNIVERSAL CHARACTERISTICS OF BEAUTY?

Even though it is not conscious or intentional, the biological drive to reproduce and pass on our genes influences our behavior at the most fundamental levels. As being attracted to another individual is typically the first step along the path to reproduction, the very qualities that one finds attractive are closely tied to the qualities needed for successful reproduction. Many features that are considered to be attractive, in the United States and around the world, are also signals of reproductive viability and overall good health. Our ancestors who were attracted to these particular traits were the ones who were more likely to survive and successfully reproduce and, thus, to pass on those genetic preferences. Attractiveness and beauty tend to go hand in hand with fertility and health, so it is not surprising that those characteristics are admired universally.

The overall goal to reproduce also provides an explanation for some of the cross-cultural and gender differences that emerge when examining which traits are considered to be attractive. Different cultures, in different environments and with differing levels of resources, have different needs for survival and reproduction, and what is considered beautiful tends to vary predictably with these differing needs, as will be discussed in Chapter 5. Similarly, men and women have different biological and social responsibilities during the reproductive process, and these differences directly correlate with the traits that tend to be preferred in a partner. These differences will be discussed throughout the text as they apply. Possibly, at some point in our evolutionary history, there were individuals who were attracted to traits other than the ones outlined in this book, but those individuals were not as successful at reproducing and passing on those traits, and therefore, we see limited evidence of such preferences in today's societies.

DECONSTRUCTING BEAUTY

Researchers examining the individual features of attractiveness typically ask participants to rate images, photographs, or individuals of the opposite sex for level of attractiveness. However, Dr. Jo Ellen Meerdink and colleagues from the University of Nebraska found that the gender of the rater does not make a significant difference in the ratings of attractiveness. For example, when asked to rate photographs for facial attractiveness, women and men tended to rely on the same set of features when rating a female face, and they relied on a different set of features when rating a male face. For both genders, face shape, eye and hair color, and hair length are considered when judging attractiveness, regardless of the gender of the rater. For female faces, the smile, eye shape, chin size, and skin texture factor in to the overall level of attractiveness. For male faces, eye spacing and size and nose size in relation to eyebrow shape and placement contribute to attractiveness ratings. Across research studies, men and women tend to rate photographs similarly, and their ratings are based on the same features, at least for Caucasian faces. It would be interesting to examine such differences when looking at faces from other races. For example, research could determine whether hair color continues to be a determining feature in Asian or African populations, considering it is more similar across individuals.

Research reveals that there are many individual traits that culminate to produce a beautiful individual. These traits include both physical and psychological characteristics, and as previously noted, most are directly tied to good physical and mental health. An individual who has strong genes, a healthy developmental environment, good nutrition, and a supportive network tends to also have the highest ratings of beauty. Thus, attractiveness is a culmination of genetic and environmental factors. Some of the predominant factors that lead to cross-cultural assessments of attractiveness include body symmetry and averageness, traits that indicate increased femininity or masculinity (sexually dimorphic traits), youth, body size and shape, scent, facial features, vocal quality, and personality traits. These characteristics, and others, are explored throughout this chapter, and the individual components are further examined in Part II.

Symmetry and Averageness

As illustrated by Michael R. Cunningham from the University of Louisville, humans all over the world share a sense of what is attractive, regardless of nationality, race, socioeconomic status, or age. Part of the explanation for this cross-cultural similarity is symmetry. Symmetry, or the degree to which the left side of the body or face matches the right side, catches the attention of the human eye. Adults around the world are more interested in and rate a symmetrical face as more attractive than an asymmetrical face. This is the underlying reason that even babies can identify a beautiful face—they are enamored by the symmetry. Interestingly, even macaque monkeys will give more attention to a symmetrical face than to an asymmetrical face of another macaque.

Although symmetry matters for both the body and the face, facial symmetry is of particular interest to researchers because we spend so much time looking at others' faces when communicating. Researchers, such as Dr. Kendra Schmid from the University of Nebraska Medical Center, used 29 different measurements to ascertain the level of facial symmetry. After identifying

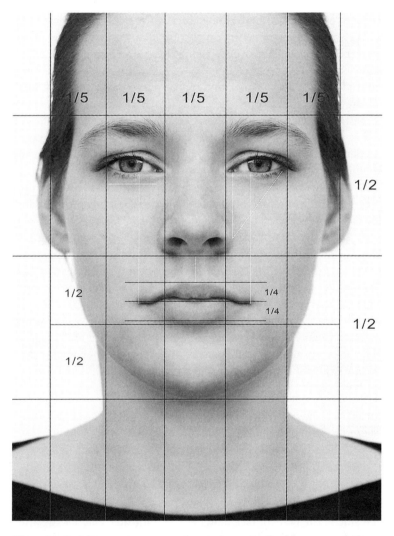

Objective facial attractiveness can be mathematically demonstrated. Standardized measurements illustrate degree of symmetry and facial proportions, which predict attractiveness. Greater symmetry is also correlated with better physical health and mental stability. (Lisavan/Dreamstime)

Defining Beauty

the most symmetrical faces, these researchers found that these symmetrical faces were the highest rated among diverse participants, regardless of the race of the individual in the photograph or the race of the rater.

In a quest for ultimate symmetry, Rotem Kowner from the Hebrew University mirrored one side of an individual's face hoping to achieve perfection. The results, surprisingly, did not reveal a perfectly beautiful face, and the perfectly symmetrical mirrored image actually received lower ratings than the unsymmetrical alternative. The reason for the decline in ratings of attractiveness was that blemishes and deviations were also mirrored. These blemishes were a signal of imperfect youth and health. However, when multiple faces were merged into one photo, or averaged, the resulting image became more and more symmetrical (individual blemishes and asymmetries were averaged and essentially eliminated), and the resulting images were rated as more and more attractive. The preference for these more average faces was cross-cultural and emerged in studies with North American, Britain, Australian, Japanese, and African hunter-gatherer participants.

Research shows that facial symmetry can be used as a generally accurate indicator of overall health and developmental stability, and average faces tend to be more symmetrical. Marked asymmetries reveal disease, genetic deficiencies, accidents, or otherwise poor health. Symmetrical faces, on the other hand, indicate good health, stable development, and high genetic quality. Thus, humans have evolved to find symmetrical faces to be more attractive and are more inclined to pursue a symmetrical individual for reproductive efforts. Average faces, similarly, denote a diverse genetic code that is made up of diverse traits that create a more attractive face. For this reason, averageness and symmetry tend to be linked because average faces tend to be more symmetrical and are therefore rated to be more attractive.

However, an interesting exception is found for the most attractive faces. Although averageness tends to be highly correlated with and predictive of the ratings of attractiveness, the most attractive faces are actually those that are not average but those that are made up of highly distinctive features. Unfortunately, having distinctive features is just as likely to make others rate an individual as highly unattractive as they are to create a uniquely attractive face. Therefore, average faces are more likely to be considered attractive than those with distinctive features, but the exact right combination of distinctive features can produce an even more attractive individual.

For example, in most studies across the field of attractiveness, females rate average male faces as more attractive than male faces with distinctive features. Composite faces (those made by blending multiple male faces) are rated as more attractive than any of the component photos. In addition to averageness, masculinity also impacts the ratings of attractiveness. Males with strong brow lines and prominent jaws are rated as more dominant, masculine, and attractive; however, this preference varies over the menstrual cycle. During the bulk of the month, women rated more average, feminized male faces as more attractive. When asked to rate the attractiveness of male faces in an experiment performed by Anthony Little and Peter Hancock from the United Kingdom, ratings of attractiveness increased as faces were averaged. Furthermore, more feminine male faces were rated more highly than masculine male faces. When looking at face shape, an average face shape was more attractive and averaging facial texture (which serves to smooth the skin) increased the ratings of femininity and attractiveness.

Body symmetry, similarly, is a contributor to attractiveness. A symmetrical body indicates healthy development and stronger reproductive potential. Asymmetries are correlated with disease, malformation, or poor genetic quality. Specifically, for women, symmetrical breasts are an indicator of sexual maturity, good health, and reproductive potential. When examining data from older women, researchers find that older women with symmetrical breasts tended to have more children over their life span than women with asymmetrical breasts. Symmetrical breasts are more attractive to the typical male, and women with symmetrical breasts have more children, on average, during their lifetimes. Research also shows that this physical symmetry is hereditary, so these women will likely pass on this trait to their offspring, contributing to reproductive success for generations to come. For males, body symmetry also contributes to increased ratings of attractiveness. However, symmetry and overall attractiveness were not found to correlate with semen quality in a study done at the University of Western Australia. Symmetrical bodies, however, did correlate with reproductive opportunities. Men with symmetrical skeletons were found to have had sex earlier and with more sexual partners throughout their reproductive years.

Symmetry, remarkably, aligns with other attractive features that will be discussed later in the chapter. Symmetrical men and women tend to have more pleasing voice quality, produce a more attractive natural body scent, have more stable personality traits, and have greater psychological and emotional health. All of these traits are likely produced by the same underlying strong genetic quality, stable environment, and healthy overall development. Based on these correlations, it is no wonder that symmetry is a hallmark of beauty because it serves as a visible physical indicator of overall physical and mental health and stability and is correlated with many other traits that contribute to future health and success.

Sexual Dimorphism

Aside from symmetry, exaggerated traits characteristic of one's own gender, such as highly feminine traits in a female or masculine traits in a male, contribute to perceived attractiveness as well. These characteristic differences

Defining Beauty

between the genders contribute to sexual dimorphism, or the distinct differences in physical appearance typical for each gender. For example, the most attractive men tend to have exaggerated masculinity: they are taller and more muscular and have broader shoulders and an inverted triangular-shaped torso. Testosterone stimulates more masculine facial features such as a solid jaw, defined brow ridge, and facial hair. The most attractive women, alternatively, have features consistent with exaggerated femininity: they are shorter on average than men and have an hourglass shape with a narrow waist and curvy hips. Estrogen inhibits masculinization of the face and instead stimulates the development of fuller lips, smaller jaws and noses, larger eyes, and defined high cheekbones.

These external sexual characteristics place a great demand on metabolic energy for one's physical development and maintenance. Furthermore, despite the myriad benefits of testosterone, it has been found to inhibit immune system functioning making men more susceptible to infections. Therefore, only the healthiest and genetically strongest individuals have the metabolic resources to create and maintain sexually attractive bodies. Only the healthiest individuals should demonstrate these extreme sexually dimorphic features, thus communicating genetic health in a similar way as symmetry.

Ultimately, sexual dimorphism has been decreasing over evolutionary time. Males and females used to have much larger differences in body size and facial features and they have become more similar over time. Although today's average female is still smaller than today's average male, there is much more overlap and similarity. The reason for the changes in overall body size over evolutionary time tends to relate to what each gender finds attractive. Males rate females with narrow waists; curvy hips; delicate facial features; clear, smooth skin; and healthy smiles as more attractive than the alternatives. These preferences have been quite consistent through history, between cultures and within individuals. Female preferences are more variable and are related to both changes in parental investment over previous generations and hormonal changes during the individual woman's menstrual cycle.

This means that changes in overall body size differences between males and females are largely tied to which men women are selecting as reproductive partners. Throughout evolutionary time, as humans have conquered and tamed our environment, women have become more interested in selecting more caring men who will invest in and help raise their children rather than merely selecting men strong enough to protect them in a dangerous environment. Today's males, thus, have an increased level of investment in child care, and this extra investment means that the children are more likely to survive.

The males who are more likely to care for children tend to have more characteristically feminine traits. Thus, today's females are more likely to be attracted to males with more stereotypically feminine traits, especially when the woman is seeking a long-term partner to father her children. These traits include being more caring, more communicative, and more physically and emotionally supportive, rather than being more muscular, dominant, or aggressive. These more feminine males, therefore, have increased reproductive fitness, meaning that they are likely to pass on their genes to more children, including the genes for these more feminine traits, leading to a decrease in sexual dimorphism between males and females over evolutionary time.

However, there are circumstances when a woman may be seeking a thrill and excitement outside of a long-term reproductive relationship. In these circumstances, she is more likely to be attracted to more masculine, dominant, and aggressive partners and is less interested in his propensity to invest in children. In this circumstance, women run the risk of conceiving a child with a man who may not stay to protect and provide for them, so the risk is great, and so, when selecting an aggressive partner, women tend to be particularly choosy. As we will discuss later, short-term relationships or one-night stands may be particularly beneficial for a male because he can potentially pass on his genes without risking resources, but a one-night stand is particularly risky for a woman because she can become pregnant without a partner to help support her and her offspring. Thus, for women, relationships become a balancing act between selecting a fertile, masculine, unreliable partner or a caring, feminine, more supportive one. Research shows that the type of relationship a woman seeks and the risk she is willing to take actually vary predictably with her menstrual cycle, and choices are made based on sexually dimorphic traits. Women at peak fertility, about two weeks after the last menses, are at the point where they are most likely to conceive a child, and prefer masculine men who carry strong masculine traits. At other points in their cycles, they prefer feminized males who will be caring, empathetic, and good providers.

Youth

Age is another major characteristic that significantly contributes to attractiveness. Age is typically directly correlated with overall health and future reproductive potential. Because true age, health, and reproductive potential are not necessarily discernable to the eye, humans must rely on a myriad of traits and characteristics to give clues. David Buss, from the University of Texas, Austin, has demonstrated through his research that physical traits associated with youth that are typically considered to be attractive include characteristics like full lips, clear skin and eyes, shiny hair, toned muscles, and high energy. These traits are also highly correlated with health, and therefore, youth and health tend to go hand in hand. Health can also be judged through the degree of symmetry and degree of sexual dimorphism (femininity vs. masculinity). As individuals age, lips tend to thin and wrinkles form,

Defining Beauty

muscles sag, energy declines, individuals start to look more gender neutral and become less fertile, and, thus, objective beauty departs. Older individuals who are considered attractive usually defy their age and look young and healthy. Alternatively, younger individuals who are lethargic, sickly, or scarred, and who look older than they are, tend not to be rated as attractive.

Youth seems to be a particularly important feature for female attractiveness. This is not unpredictable because youth is tied to reproductive potential, particularly for women. Thus, it makes sense that youth cues are more important to men than to women when seeking a reproductive partner. Immature facial features that indicate youth, such as a small nose, full lips, and large eyes, are particularly alluring to most men. Models tend to have more immature features on average than other women of their same age, partially explaining why they are considered to be so attractive. Being attracted to a more youthful woman increases the chances of successful reproduction for a man. Thus, men who are attracted to such traits tend to have more children and pass this tendency to the next generation throughout evolutionary time.

To examine the impact of different physical features on the ratings of age and health and the impact each has on the ratings of attractiveness, Devendra Singh and colleagues from the University of Texas, Austin, had participants rate line drawings of women. Those women who were perceived to be younger were perceived to be more attractive. The drawings varied on overall body size (body mass index or BMI), degree of hourglass shape (i.e., waist-to-hip ratio or WHR), and breast size. When asked to estimate the age of each woman, the participants used these three characteristics, body weight, WHR, and breast size, in a similar manner to make their judgments. Those drawings that indicated more slender figures, lower WHR (more hourglass figure), and smaller breast size were rated to be younger. Also, larger figures with low WHR were perceived as younger than larger figures with a more tubular shape. Slender figures with large breasts were judged as slightly older than those with small breasts. When asked about attraction and health, the slender figures with low WHR ratios and either size breasts were uniformly rated as the youngest, the most attractive, the healthiest, and the most feminine looking.

Doug Jones, a visiting scholar at Cornell University, examined correlations of age and attractiveness around the world. He analyzed photographs of women from the United States, Brazil, Russia, and the native populations of the Ache Indians and the Hiwi. Facial proportions were measured for all of the photographs, and as expected, the proportions changed predictably with age. Due to individual variation, however, some of the women had proportions that were typical of younger women, and some women had features that already appeared more mature than their actual age. Across the five cultures, the correlation of age and attractiveness was examined. Females, across all the cultures, were rated as more attractive if their predicted ages, based on the facial proportions, were less than their actual ages. Thus, ratings of attractiveness were directly rated to perceptions of age based on facial features. When the same kind of procedure was repeated for males, the same correlations were not found or were very weak. Thus, age is not as valued a predictor for male attractiveness as it is for females.

Body Mass Index

BMI is calculated based on an individual's overall body weight in relation to his or her height. BMI has long been thought to be an important cue in assessing attractiveness and health. Based on decades of research, BMI seems to be a more important factor for female attractiveness and beauty than for male attractiveness. Women with higher BMI tend to be rated as less attractive in research done in the United States and other industrialized countries than women with lower BMI. Furthermore, lower BMI tends to be correlated with models, supermodels, and actors and tends to be rated as more attractive, at least in industrialized societies.

However, the influence of BMI on ratings of attractiveness varies between cultures. Industrialized societies correlate lower weight with health and reproductive potential and link it to attractiveness. Nonindustrialized societies, however, are more tolerant in their ratings of larger women. This cultural difference conflicts with the ideal of a universal idea of beauty, but the differences make logical sense when examined through an anthropological or social lens. In nonindustrialized societies, with limited or less stable food resources, larger women tend to be healthier, hardier, and more likely to survive during the reproductive process. Thus, in societies like Kenya, Uganda, and Tanzania, men predictably rate larger women as more attractive and more desirable. They also rate those women as having more potential as spouses and reproductive partners. Ultimately, less industrialized cultures, with less stable food resources, were more likely to rate larger women as more attractive than slimmer women. Alternatively, more industrialized cultures, with greater access to stable food resources, had ratings more similar to the low BMI preferences typical of individuals from the United States.

In addition to the cross-cultural differences, preferences for ideal BMI vary within countries as well. Within the United States, for example, African Americans are more likely to rate larger, curvier, women as more attractive, whereas Caucasian Americans do not rate those women as highly. These differences seem to be a by-product of culture and of the media rather than of biology. As a side note, these differences in ratings by culture have direct

Defining Beauty

implications for mental health, and cultural preferences for low BMI are correlated with prevalence of eating disorders such as anorexia and bulimia, with Caucasian girls being at particular risk for developing eating disorders. African American girls do not tend to have the same cultural pressure to be underweight and therefore are more likely to embrace their healthy developing bodies more openly and with healthier attitudes.

The impact of BMI on male attractiveness does not seem to play such a large role cross-culturally. Females seem to be more interested in body shape rather than body size. Females rate larger, more muscular, males as more attractive cross-culturally. Although extremes in weight (extremely over- or underweight) impact ratings, average differences in overall body weight did not contribute to ratings of attractiveness as much as they do when either gender is rating for female attractiveness.

Body Shape: Waist-to-Hip Ratio and Shoulder-to-Hip Ratio

Renowned researcher Devendra Singh and colleagues from the University of Texas, Austin, have been studying the impact body shape has on attractiveness for over 20 years. Dr. Singh's research focused on WHR in women, likely a precursor to the more recent shoulder-to-hip ratio (SHR) research in men.

WHR is the comparison of the circumference of the waist to the circumference of the hips. A low WHR (e.g., 26-inch waist and 37-inch hips have a WHR = 26/37 = 0.70) creates the stereotypical hourglass feminine figure. A high WHR (e.g., 30-inch waist and 33-inch hips have a WHR = 30/33 = 0.91) creates a more tubular body shape that is more common in prepubescent children, postmenopausal women, and males.

Women's WHR typically ranges from 0.67 to 0.80 and men's WHR ranges from 0.85 to 0.95, meaning that the WHR between typical, healthy females and males do not even overlap. The most attractive WHR in women, as rated by observers, ranges from 0.68 to 0.72. When examined, researchers found that historical Greek, Roman, Egyptian, Indian, and Japanese statues and clay figurines had WHR between 0.63 and 0.69. Currently, most models, Miss America winners, and *Playboy* centerfolds have a WHR within the 0.68–0.72 range. Even Twiggy, the English model from the 1960s who was famous for her slender and androgynous build, had a WHR of 0.73, just slightly outside the typical ideal and still lower than average for an average woman. And, Marilyn Monroe, American actress and model heralded for her voluptuous sexuality, had a WHR around 0.70, making her the ideal hourglass shape. Although the overall weight of models has been decreasing over the decades, the ideal WHR has remained the same. As the WHR is a comparison of hips to waist, overall body size or BMI is not considered, just the difference in the waist circumference and the hip circumference regardless of overall body size.

When examining the impact of the comparison between the waist and hips on attractiveness, WHR has emerged as a more accurate assessment of attractiveness and of health than BMI in psychological studies. Even in those less industrialized cultures, which rate larger women as more attractive, the WHR preference remains the same as that found in industrialized cultures, which revere thinner women. Studies completed in North America, Europe, Africa, Indonesia, and New Zealand revealed a similar relationship between WHR and attractiveness ratings, even when overall body size was experimentally manipulated. Thus, the shape of the body seems to be more universally relevant than overall body size. Furthermore, WHR is arguably a more reliable indicator of overall health than BMI.

WHR is a defining feature of sexual dimorphism between the sexes, at least during the reproductive years. Before puberty, girls and boys have similar body shapes. After puberty, the average WHR between men and women in their reproductive years do not even overlap. The differences in WHR between the genders result from fat distribution during puberty, which is regulated by sex hormones. During and after puberty, women tend to gain around 35 pounds of fat around the hips and thighs. This decreases the WHR by increasing the circumference of the hips and creating the feminine hourglass figure. Post menopause, deposits of fat around the midsection increases the WHR, making the woman's body more similar to a male's and decreasing the ratings of attractiveness. This demonstrates that WHR has direct implications for and visually communicates reproductive value and fertility. When a female has a higher, more masculine WHR (e.g., as found in prepubescent girls and postmenopausal women) this indicates she is not at peak fertility and has lower immediate reproductive potential. Accordingly, others rate these individuals as less physically attractive.

While BMI and WHR do not seem to play a large role in ratings of male attractiveness, an ideal male's WHR is around 0.90. This creates a tubular shape of the lower body. Variations around this ideal, however, do not significantly impact ratings of attractiveness for males. A trait that emerges as more indicative of attractiveness ratings for the male form is the SHR, which has been more recently studied as more predictive of attractiveness in males than either WHR or BMI. In differing research studies, the male shape has also been studied as the waist-to-shoulder ratio or waist-to-chest ratio. For ease of discussion, SHR will be used to discuss this feature here.

As mentioned, SHR emerges as more important than overall BMI when men and women are asked to rate male attractiveness. SHR compares the circumference of the shoulder/chest to the circumference of the hips/waist in males. A male with a high SHR (50-inch shoulder and 34-inch hips would have an SHR = 50/34 = 1.5), who has broad shoulders and a narrow waist or hips, is universally found to be more attractive, healthier, and more masculine, and perceived as more dominant than males with a lower ratio. A male with a low SHR (40-inch shoulders and 38-inch hips would have an SHR = 40/38 = 1.05), who has shoulders and hips that are about the same size, is rated as less dominant, lower in physical attractiveness, and less masculine.

Research on SHR has revealed that men with a high SHR do tend to have stronger immune systems, and are more masculine, healthier, more competitive, and more physically powerful than men with lower SHR. To gather empirical evidence that this trait actually impacts attractiveness ratings, Barnaby Dixson, from the University of New South Wales, and colleagues assessed where and how long women tend to look when presented with a nude male standing in a back pose. Women in their study rated men with higher SHR as more attractive, and spent more time shifting their gaze between the man's upper and lower back. For photographs of heavier or scrawnier men, with lower SHR, the lower back captured more of the women's eye gaze, rather than the shoulders. This study demonstrates that, even at a behavioral level, women's attention is drawn to those body parts that reveal a male's underlying genetic health. Men with a high SHR tend to be healthier, and have better cardiac function, greater strength, and a stronger immune system, all of which are arguably attractive to a potential partner. Based on where women naturally tend to shift their gaze when looking at these men, they are unconsciously taking in all of this information. Men with lower SHR are likely to be less healthy, have more body fat or less muscle, and are more susceptible to disease. All this visual information is subtly revealed in their body type and leads women to rate them with lower physical attractiveness scores.

SHR has not been researched to the same extent as WHR, but this existing research does show that it serves as a marker for females when assessing the physical attractiveness of a male. Testosterone levels impact the distribution of body fat for males just as estrogen impacts the distribution for females. In males, testosterone exposure creates the inverted triangular shape of broad shoulders and a narrow waist. Body shape, thus, is a clear indicator of underlying physiological and hormonal functioning and is a clear visual cue for these unobservable traits. Thus, men and women are attracted to particular features of body shape even when we are not consciously aware of why we find such traits attractive. In both genders, body shape seems to be a more accurate indicator of attractiveness than overall body mass, despite the initial claims by research on BMI.

Body Scent

Pheromones, or the scented chemicals that are released when an individual sweats, are just starting to be understood more fully in relation to human attraction and attractiveness. Pheromones are known to play a role in animal courtship, but their role in human attraction is not as well understood. The influence of pheromones takes place at an unconscious level, perhaps directing one's attention toward another individual and potentially contributing to a sense of instant chemistry with another person. But conscious decision making and interpretation of these unconscious signals definitely play a significant role in ultimate behavior and assessment. Research shows that the chemical signals we emit actually influence the hormone levels of other people. Pheromones play a communicative role in many other species, and it is likely that they have a function in human relationships as well.

A woman's scent and her sensitivity to the scent of others change over the menstrual cycle. In women, pheromones are secreted from the genital area, the area around the navel, the breasts, and the armpits. If these scents are too strong, they can create an aversive effect in the form of body odor, but, at low quantities, they may be unconsciously enticing. In this way, pheromones can have a regulating effect on the hormones of those around them. A specific example is the phenomenon of the synchrony of the menstrual cycles of women who are frequently together. James Kohl from JVK Resources, Inc., along with researchers from the University of Vienna demonstrated the effects of this hormone exposure. By exposing participants to artificial pheromones, they were able to significantly increase or decrease reported sexual attraction in participants, synchronize menstrual cycles, and alter mood, as compared to control groups.

Body scent is also correlated with genetics, health, and immune system's functioning. In general, a woman tends to prefer the smell of a man whose genes vary from her own. If reproduction were to occur, this variance would create diverse offspring with varied immune systems. In a genetic sense, mating with a sibling would create redundancy and thus most siblings do not smell attractive to one another. In fact, they tend to smell repulsive.

Women tend to be more sensitive to smells than men and, thus, may be more apt to detect and react to scents of an attractive (or unattractive) male. During ovulation, women are the most sensitive. Markus Rantala and colleagues from the University of Jyvaskyla and the National Public Health Institute in Finland found that women, particularly those who were not using contraceptives, tended to rate masculine scents as more attractive while in the most fertile phase of their menstrual cycle. To examine the usefulness of smell for detection of attractiveness, Steven W. Gangestad and Randy Thornhill from the University of New Mexico asked women to smell and

Defining Beauty 17

rate sweaty T-shirts worn by males. The scents that women rated as most attractive in the blind study were significantly more likely to belong to more symmetrical, healthier, more objectively attractive men. Symmetry, a well-known factor in the level of attractiveness, apparently exhibits itself through scent as well.

Although women tend to be more sensitive to scent, men also seem to be able to use odors to detect a woman's level of fertility. In a similar study, men were asked to rate sleep shirts worn by women. Findings revealed that the scents of those women at the peak of fertility (during their monthly cycle) were rated as most sexy and pleasant. Similarly, when men were asked to rate photographs of women who were in different phases of their cycle, those women at peak fertility were rated as most attractive. In both genders, attractiveness of scent was tied to physical attractiveness and reproductive value, strengthening the importance of this biological signal as a contributor to beauty. In both genders, however, scent did not outweigh visible physical cues, and it remains to be a subtle and unconscious force, likely directing attention but not determining ultimate decisions or behavior.

Vocal Quality

Examining another sense, research also shows that vocal quality is directly correlated with the level of attractiveness. Vocal pitch tends to differ greatly between the sexes, and women rate lower-pitched male voices as more attractive, while men rate higher-pitched female voices as more attractive. In a psychological study, participants rated voices for the level of attractiveness and when the rated voices were matched up to the individuals, research revealed that the higher-rated voices belonged to individuals who were physically more symmetrical and who had greater physical sexual dimorphism than those in the lower-rated groups. Additionally, men with lower-pitched voices tended to have a greater SHR with broader shoulders and narrow waists, and higher-pitched females tended to have low WHR with small waists and curvy hips. These connections are not that surprising because vocal quality, symmetry, sexual dimorphism, and scent are all influenced by hormone exposure during puberty, which starts this cascade of fertile and attractive development.

Personality

Although physical characteristics are highly influential in explaining attractiveness, other nonphysical factors have been found to be highly influential as well. Personality characteristics have emerged as highly impactful on perceived level of overall physical attractiveness. Beyond physical traits, many personality and social traits significantly contribute to the ratings of physical attractiveness. Personality may be an even more important trait to women than physical traits when assessing a man's overall attractiveness. Women tend to be more influenced by indicators or clues about a man's level of ambition, kindness, and empathy than they are by his symmetry and youth. When anthropologist Dr. Heather T. Remoff asked women to rate photographs of males for level of attractiveness, women were more likely to rate males who were kindly interacting with children as more attractive, regardless of their actual physical traits. They found these males more attractive than males on their own or males posing (but not interacting) with children. Men interacting with puppies and kittens tended to elicit a similar positive reaction, even when the women were specifically asked to rate physical attractiveness.

Viewing such an interaction between a man and a child may indicate to a woman an ability to provide for a future family. Such interactions may also insinuate that the man is kind, compassionate, good with children, and committed to family. Interestingly, even women who were not ready or not planning to have children rated these men as more attractive. The indicators that such an interaction gives about the man's traits go beyond just child care and demonstrate a potential openness, kindness, and level of empathy that he will likely bring to the relationship regardless of whether or not children are produced. Keep in mind that women were asked to rate for physical attractiveness, not necessarily for mate value or father potential. The man's behavior actually increased his perceived physical attractiveness in the eyes of the female raters.

When examining personality, many researchers have used the Big-Five model of personality to examine the effects of personality. The Big-Five model of personality suggests that an individual's personality varies along five dimensions. These are level of extraversion, degree of openness to new experiences, level of conscientiousness, level of agreeableness, and level of neuroticism.

Individuals who are high in extraversion are those who are sociable and outgoing and demonstrate interpersonal confidence. Extraversion is tied to the perception of increased social experience, heightened social skills, and more social confidence. Extraverts are likely to be adventurous and have more experiences with dating and sex, making them more likely to successfully reproduce than introverts. Extraverts are frequently rated as more attractive than introverts who are less likely to take risks, less likely to engage others, and less likely to have social experience and confidence.

Individuals who are high on openness to new experiences are those who have a high level of creativity and imagination, and demonstrate an open

Defining Beauty

curiousness about the world and others. Individuals who rate high on openness to new experiences tend to be rated as more attractive and may be innovators and be open to change. They likely have more social relationships as well as more sexual experiences. Closed individuals tend to be rated as less attractive by others and are more cautious, less likely to engage others, and less likely to seek out new experiences.

Individuals who are high on conscientiousness are orderly, hardworking, and attentive to details. Conscientious individuals have higher self-control and tend to be more achievement oriented. They pay attention to details in work and relationships and thus are more attractive to others for romantic or professional pursuits. Research has also noted that conscientious individuals tend to be more physically symmetrical, which may also contribute to their higher ratings of attractiveness.

Individuals who are high on the dimension of agreeableness are kind, empathetic, and trusting. Individuals with high levels of agreeableness are consistently rated as more attractive. Dr. Jensen-Campbell from the University of Texas, Arlington, designed a series of early studies demonstrating the importance of an agreeable or prosocial orientation on the ratings of attractiveness. In this study, agreeableness emerged as a highly attractive trait. Although masculine men would be predicted to be the most attractive, agreeableness emerged as even more influential. While masculine men, who were also reported to be high in agreeableness, were rated as the most attractive, level of agreeableness emerged to be six times more important than level of dominance and masculinity. Even men who were only average in level of attractiveness, but who were described as agreeable, were rated as more physically attractive than objectively attractive men who were described as disagreeable. Agreeableness also had a greater impact on attractiveness ratings than physical traits such as SHR. Agreeableness was so influential on perception of attractiveness that women actually rated disagreeable men as unattractive, even when they were described and physically appeared to have a high level of masculinity and symmetry. Later research from Dr. Jensen-Campbell's lab demonstrated a similar effect on men's ratings of women's attractiveness. Agreeable women were rated as more physically attractive than less agreeable women. Agreeable people tend to be more sensitive to others' moods and more likely to have larger friend networks, while those with low scores on agreeableness tend to be considered egocentric and selfish.

Individuals who are high on neuroticism are more likely to be anxious, depressed, and susceptible to feeling threatened. Those with low neuroticism are stable, well adjusted, and comfortable. Highly neurotic individuals are less likely to have healthy, stable relationships and tend to be rated as less physically attractive. People with high scores on neuroticism tend to be wary of others and are more sensitive to changes in social situations. Individuals who are low in levels of neuroticism are likely to be less anxious, but may also be less sensitive to social changes and less perceptive about the feelings of others, making the level of neuroticism less consistent as a predictor of attractiveness.

So, testosterone and measureable physical traits are attractive, but alone, they are not enough. Personality matters. Information about another's personality along dimensions of extraversion, openness, conscientiousness, agreeableness, and neuroticism can actually make another person seem more or less physically attractive, even without altering his or her physical appearance.

The Big-Five factors of personality are relatively independent from one another, and these personality traits tend to remain relatively stable throughout adulthood. There are typical gender differences between the dimensions, including that women tend to be more extraverted, more conscientious, more agreeable, and more neurotic than men. Although research predominantly discussed how personality might affect one's perceived level of attractiveness, it is interesting to note that attractiveness can also affect one's perceived personality. Attractive people tend to be assessed more positively on traits beyond just physical appeal and tend to be labeled as more outgoing, less neurotic, and more open, even before adequate information is known about the individual. Thus, first impressions regarding another person's personality are made, for better or worse, setting up expectations about future behavior based on physical traits alone.

Research demonstrates that more physically attractive individuals are perceived as more extraverted, particularly for male targets. Attractive men were also rated as more open to new experiences. They were perceived as more curious and imaginative, even when the raters only had a photograph on which to base their expectations. Photographs of attractive men and women were perceived to be less neurotic, and attractive women were rated to be more conscientious. The level of physical attractiveness did not influence how agreeable a person was perceived to be, although research does show that the level of agreeableness does influence the ratings of attractiveness, as previously discussed. For all of these dimensions, the stereotypes associated with attractiveness influenced the ratings of personality traits. More positive personality traits were attributed to physically attractive individuals, although these expectations differed between the genders.

One could argue that just because someone is rated as more extraverted or open based on their physical appearance, it does not mean they actually are. To address this connection and to examine the veracity of first impressions, Dr. Laura Naumann and colleagues from the University of California asked participants to rate personality qualities based on photographs of individuals.

Once the ratings were collected, they were compared to the photographed individuals' actual personality traits. The results revealed that the ratings for agreeableness, stability, openness, likability, and loneliness, made without information about the person in the photo, were actually highly accurate. Perhaps the way in which the individuals held themselves in the photos was highly revealing of their actual traits.

Other personality traits that are not captured in the Big-Five model, such as interpersonal interest (caring, asking questions, listening), cooperation, honesty, helpfulness, empathy, and selflessness, were also found to be traits that increased the ratings of physical attractiveness. Furthermore, a sense of familiarity, mutual respect, and signs of motivation and ability increased the perceived physical allure. A sense of camaraderie through shared adventures or a sense of intimacy from shared experiences and conversation also increased attractiveness ratings. Being perceived as helpful, honest, and considerate of others also boosts the attractiveness ratings. At times, these intangible features were found to contribute more to physical attractiveness ratings than did the person's actual physical traits, particularly when the participants were asked to envision having a friendship or romantic relationship with the person in the photos they were evaluating. A person described as negative, rude, or selfish was perceived as less attractive, even though that person's physical features were unchanged.

In an interesting study that clearly illustrates the direct link between attractiveness and personality, men and women were asked to rate photographs of the opposite sex with and without information about personality. They first ranked the photos without any additional information and then they were asked to make ratings of photographs after being given descriptions that included either positive or negative personality traits. When the raters had information about the personality of the individual in the photograph, their ratings of level of attractiveness changed significantly. Interestingly, negative information about personality traits had a bigger impact on the ratings of attractiveness than positive information. Furthermore, information about personality had a more substantial effect on ratings of attractiveness than information about WHR or body shape. For positive personality characteristics such as extraversion, agreeableness, conscientiousness, openness, and stability, a broader range of individuals were rated as attractive. For negative characteristics, only the most physically attractive individuals were still rated as attractive. Ratings were the most improved for individuals with at least average attractiveness. Having positive personality traits was not as helpful for extremely unattractive individuals but it was highly influential for individuals in the average or above-average ranges. These results of positive personality descriptions increasing the ratings of attractiveness and negative personality descriptions decreasing the ratings were found in multiple studies. Remarkably, but not altogether unsurprisingly, women were particularly adverse to and influenced by descriptions of negative personality traits when rating others.

In a literature review of a large body of research on this topic, Domagoj Švegar from the University of Rijeka analyzed empirical findings that examined the relationship between symmetry and personality. Symmetry, as we discussed earlier, is a by-product of genetic potential and the stability of the developmental environment. Good genetic quality provides resistance to environmental stress, poor nutrition, and toxins, while low stress, good nutrition, and low toxin exposure can help maximize genetic expression. Symmetry, particularly facial symmetry, is one of the key markers that individuals use when estimating the level of attractiveness.

Dr. Švegar recognized that one of the struggles of studying the relationship between attractiveness and personality is whether facial symmetry predicts particular personality traits or whether facial symmetry evokes particular responses from others, which in turn elicits particular personality traits from the individual. As personality is arguably a sum of genetic potential and environmental experiences during development, how one is treated will likely affect his or her ultimate personality. An attractive child will likely be treated more positively and kindly than an unattractive child, which will have corresponding effects on the child's later personality. So, the link between personality and physical attractiveness may be excessively intertwined and partially explained by one's early development. An attractive individual may have a higher likelihood of developing a healthy and stable personality due to more positive environmental interactions, while an unattractive individual may encounter more of a struggle.

Dr. Švegar's analysis of the research on symmetry demonstrates that symmetrical individuals are healthier than asymmetrical individuals, although the actual difference in the level of health is less than what is typically perceived (i.e., symmetrical individuals are rated as much healthier but, when analyzed, are only slightly healthier). However, given that they are, on average, physically healthier than asymmetrical individuals, the question becomes whether or not they are also psychologically healthier than asymmetrical individuals. Shackelford and Larsen found that symmetrical individuals were rated as less neurotic, less emotional, less angry, less anxious, and more agreeable and conscientious than asymmetrical individuals. Asymmetrical individuals, alternatively, were rated as more anxious, less balanced, less intelligent, less confident, and less sociable. Again, it is difficult to say whether these traits result from social interactions throughout development or are projected onto the individual due to his or her physical appearance.

In this body of research, ratings of personality traits were consistent across studies. This shows that individuals draw conclusions about others' personality traits based on symmetry. Researchers, then, shifted their focus to analyze whether or not the personality ratings based on symmetry are accurate. Just because someone is perceived as neurotic or agreeable based on his or her level of symmetry, it does not necessarily indicate that the individual actually possesses those traits. Interestingly, when looking at actual traits, some research demonstrated that the patterns of findings shift. Symmetrical individuals were found to be low on ratings of neuroticism as predicted but they were not found to be more agreeable when rated by personality tests. Also, conscientiousness emerged as unrelated to symmetry, and symmetrical individuals were actually found to be less open in some research findings. Finally, extraversion was positively related to symmetry when actually measuring personality traits despite the fact that there was no such correlation when studying perceptions. Later research replicated the connection between extraversion and symmetry but found no relationship between symmetry and the other four personality dimensions.

So, raters judge individuals with highly symmetrical faces more positively and conclusions are drawn regarding the person's personality characteristics based on that person's level of symmetry. This likely creates expectations by which future behavior will be judged. So, having a symmetrical face already gives the individual an advantage before he or she has engaged in any type of conversation or activity, regardless of his or her actual personality traits.

Recent research has proposed that facial symmetry does influence personality development. Highly symmetrical individuals can be viewed as having higher status and higher value and thus they do not need to compensate to earn status. Asymmetrical individuals may need to work harder to achieve interpersonal relationships and thus may have more experience with social skills. However, when 200 personality traits were examined, a negative correlation was found between the level of symmetry and myriad positive personality qualities. In that particular study, symmetrical people were found to be less friendly, less responsible, less trustworthy, and less calm, among many other traits. Alternatively, symmetry was positively correlated with negative qualities like hostility, aggression, depression, and anxiety. Furthermore, high symmetry was related to neuroticism and asymmetry was correlated with agreeableness.

So, based on decades of research, results remain contradictory, but what is consistent is the idea that we make assumptions about others based on the way they look and those assumptions likely bias our interpretations of behaviors. Findings regarding perceived psychological health as related to symmetry were less contradictory than findings connecting actual health to symmetry. It is still unclear whether the level of symmetry influences personality or whether it is merely the perception of symmetry that creates the expectation of and ratings of certain personality traits.

While first impressions are frequently based on physical appearance, opinions can change when more information is learned. Frequently people become significantly more (or less) attractive as their personality starts to shine through and relationships extend beyond the surface. An honest, outgoing, open individual will become more attractive, even if the attractiveness actually lies beneath the surface.

Similarity, Familiarity, and Propinquity

Taking the social psychological point of view into consideration, most individuals are attracted to others who have similarities with themselves. These similarities may include level of attractiveness, attitudes, beliefs, hobbies, political orientations, intelligence, personality, social class, age, gender, race, religion, profession, and a host of other traits. Similarities are validating. If others look or think in a similar way or hold similar attitudes and agree, that is reinforcing and we tend to like them better. We are reminded of ourselves when we see the other and enjoy spending time with him or her. When others agree with us, it is validating and makes us feel good. Similarities create smoother interactions and more positive expectations and experiences.

A couple who enjoy spending time together tend to rate each other as increasingly attractive. Sharing experiences with someone who shares common interests and attitudes creates a sense of intimacy, which impacts the perception of attractiveness independently of actual physical traits. (Michael Zhang/Dreamstime)

Because we learn about what is normal from our own experiences, what we typically see becomes our own normal. In the United States, we certainly see models and actors, so our appreciation for objective, close-to-perfection beauty is heightened, but we even more frequently see our friends and families and we tend to prefer those who look like those people we know. The adage "opposites attract" is not supported by scientific evidence. Although it may be easy to think of exceptions and identify couples who have characteristics that are total opposites of one another, research shows that, the majority of the time, the pull of attraction is toward someone similar to oneself.

When someone is similar, it rings of familiarity. Research shows that repeated exposure to something novel increases one's enjoyment (at least to a point). For example, increased exposure to a song or painting correlated with higher enjoyment ratings. Familiarity and exposure can impact the ratings of the attractiveness of other people as well. The more familiar a person is, or the more a person is perceived to be familiar by looking or acting like someone already known, the higher the ratings of attractiveness. People find comfort in the familiar and are likely to rate it as more likable and more attractive. This demonstrates that beauty certainly can be in the eye of the beholder. In this sense, beauty is utterly subjective, and physical proportion and objective measures no longer hold such a strong influence over the perception as the person becomes more familiar. A familiar face becomes comforting and is rated as more and more attractive even if it does not become more objectively beautiful.

Familiarity can be earned or merely perceived. When individuals meet someone who reminds them of someone who was kind to them or who they enjoyed spending time with in the past, their ratings of attractiveness tend to increase (or decrease if the relationship was not positive). For example, teachers or professors who encounter students who remind them of previous students tend to alter their ratings based on the quality of the previous relationship. Similarly, students who encounter a professor who reminds them of a previous professor alter their ratings accordingly. Meeting a new acquaintance who has qualities that are similar to those of good friends or previous positive relationships can create a circumstance of instant bonding. Perceived similarity of physical features, voice qualities, similar gestures, or similar attitudes can bias ratings for even novel individuals.

Perhaps tied to familiarity and similarity, humans also tend to be attracted to those whom they are close to in physical space. The concept of propinquity, or of being in close proximity to someone else, seems to be a key variable in attraction and ratings of attractiveness. Since the 1950s, psychology research has demonstrated that individuals tend to marry someone who lives within just a few blocks of themselves, potentially explained by the mere exposure effect, or the tendency of individuals to like new stimuli more with repeated exposure. Being close in physical space is a key variable for meeting, becoming familiar, sharing experiences, building similarities, and cultivating attraction. Even in college dorms, friends tend to form between those individuals who live closer to one another—the same hallway, the same dorm, or the same classes. Propinquity can be created through dormitories, neighborhoods, volunteer or professional organizations, or seating arrangements. This proximity can heighten positive ratings if there are also similarities between the individuals. If there are no similarities, the proximity can have the opposite effect and dampen ratings and cause problems. Propinquity is likely effective for a whole host of reasons. Because those who are nearby are available for interaction when desired, there is a high likelihood of frequent interaction. Due to this, people are more likely to be on their best behavior to cultivate a good impression. Also, when we frequently meet a person and spend time with him or her, we get used to that person and are able to more accurately anticipate that person's behaviors, quirks, and moods, which may facilitate successful interactions.

Within the similarity and familiarity research, a threshold does emerge. Being too familiar or too similar has a negative impact of attractiveness ratings. Hence, siblings and children raised together tend to find each other less attractive. This may be an evolved safeguard against incest. Children raised together, even if they are not genetically related, tend to rate one another as less physically attractive. This can have negative effects in cultures where children are paired for future arranged marriages too early. Sending the future bride to live with the promised-husband's family at a young age decreases the level of physical attraction for both parties and tends to decrease the likelihood of reproductive success in the future marriage. So, humans like enough similarity and familiarity to create the feeling that another individual is close, but not too close!

The impact of the Internet age, or electronic propinquity, on this driving force behind attraction is currently being studied. The question is whether or not being close in cyberspace will have the same effect as being close in physical space. Certainly, individuals can now meet, talk, become familiar, and find those who share similarities from a pool of partners all over the world rather than just in one neighborhood. This expansion of the possibilities has already contributed to a delay in the age of marriage due to the vast number of potential partners that one now has to select from.

PERSPECTIVES ON ATTRACTIVENESS

As illustrated in this chapter, there are different ways to consider beauty. From an evolutionary perspective, beauty may be the result of past environmental pressures and may be a visible indicator of underlying genetic quality.

Defining Beauty

Cross-culturally, beauty is recognized in terms of symmetry, averageness, and extent of sexual dimorphism, all constructed from the collaboration between one's genetic recipe and one's developmental environment. The importance of attractiveness as a signal for underlying genetic health is particularly salient in environments without modern medical care. In such environments, signs of attractiveness are more closely tied to likelihood of survival.

Taking into account cross-cultural differences in what is considered to be beautiful, there is also a sociocultural perspective that contributes to understanding attractiveness that extends beyond genetics, which will be discussed more fully later in this book. Body mass and self-perceptions of beauty vary between cultures. The cultures that more closely adhere to traditional sex roles are the ones that prefer more traditionally attractive features and traits. Obesity is one trait that is influenced by sociocultural ideals. In affluent countries, obesity is associated with poor self-control, low self-discipline, and low morals. This is particularly troubling because some research shows that individuals who are moderately overweight live longer than those who are average or underweight. In less affluent countries, obesity is tied to wealth and status and is thus correlated with perceived attractiveness in a direct reversal of the pattern.

The perception of beauty also varies among individuals. Length of relationships, quality of intimacy and shared experiences, and age of the perceiver can contribute to the perception of attractiveness. Furthermore, feedback from others can influence one's perception of his or her own level of attractiveness. Women tend to be able to more accurately assess their own level of attractiveness, likely due to feedback from men throughout social interactions.

The biological perspective illustrates that, when perceiving an attractive individual, there is actually a resulting impact on neural activity and cognitive processing. The ventral occipital region (in the visual cortex at the back of the brain) is specifically activated when presented with an image of a beautiful individual. Different situations elicit variations in the way we process information. For example, attractive individuals (particularly attractive women) tend to stand out, are more recognizable, and are more memorable to others. Also, being in a committed relationship changes the way women perceive and process information about other attractive males. After viewing photographs of other attractive women, a woman is more likely to have a decrease in self-esteem and rate herself more poorly as a potential marriage partner. And, men who view images of other males were most affected by the reported social status of the other male. This creates a similar threat to self-esteem and self-assessment.

There may also be features internal to the individual that influence that person's perceptions of beauty. Hormone levels, fertility, and one's own personality and level of attractiveness may influence what one seeks and likes in others. For example, when women are in the most fertile of their menstrual cycles, they rate more masculine faces as more attractive. However, when they are in a less-fertile phase, they prefer more feminized faces. Masculine features such as large muscles, strong brows and jaws, deeper voices, v-shaped bodies, and quality of body odor may advertise dominance, genetic quality, virility, and good health, all qualities that a woman may choose for offspring. Less masculine features may advertise more feminine traits like increased likelihood of long-term commitment and parental investment, which a woman would likely choose for a long-term partner or co-parent.

From a social psychological perspective, beauty also tends to create a self-fulfilling prophecy, which will be explored in the next chapter. Beautiful people tend to feel better about themselves, tend to be perceived better by others, receive more help from others, make more money than others, and are given more opportunities by others. Unattractive people, particularly males, are more likely to become alcoholics, attain lower levels of education, have fewer romantic partners, be more depressed, and have more health problems. Unattractive men have fewer children, make less money, and are more discriminated against. The link between attractiveness and success seems to be a two-way street. Attractiveness provides opportunities for success and success increases the ratings of attractiveness.

The media also plays an important social psychological role in constructing beauty ideals. Beauty is represented differently cross-culturally and across time. Westernized advertisements are rated as more body oriented than advertisements in Asian countries. Research does show that there are increases in similarity of advertisements across culture with time as cultures influence one another. Hollywood films also selectively connect attractive actors with particular personality traits. Beautiful characters are more likely to be outgoing, honest, moral, and good than less attractive characters. This perpetuates the "beauty is good" stereotype. Furthermore, across the second half of the 20th century, typical fashion models have become increasingly thin and full body photographs have become more common. A similar trend emerges when examining *Playboy* centerfolds, Miss America Pageant contestants, and fashion models. These models have historically weighed less than the average American woman, but by the end of the 20th century, they weighed up to 23 percent less on average. Another trend emerges when examining how the ideal male form has shifted over time. Muscle size on GI Joe dolls gradually increased until it reached unrealistic proportions by the mid-1990s.

One's own level of attractiveness can also influence one's ratings of others. Women, for example, who rate themselves as more attractive, are more likely to be attracted to more masculine, more symmetrical, and overall more attractive males. Women who were showed pictures of highly attractive women then viewed themselves as less attractive and were less likely to prefer more masculine features in a partner. Alternatively, women who viewed photos of unattractive women viewed themselves as more attractive and were more drawn to highly masculine men. Thus, ratings of attractiveness can be subjective, even within an individual, contingent upon one's perception of his or her own attractiveness.

CONCLUSION

Psychological, biological, anthropological, and social research demonstrates that there are universal characteristics that are perceived as beautiful. Although we are not likely aware of the individual features when we see something we like, a scientific examination reveals that there are some underlying qualities that culminate in objective beauty. Upon examination, the traits that we tend to find attractive are very logical. They are those features that indicate strong genetic quality, a sound environment, developmental stability, and likely future success. Many of the features are physical manifestations of genetic qualities that will also lead to strong, healthy offspring. Additional traits that contribute to attractiveness, like personality and previous experience, only add to the possibility of successful, long-term, enjoyable relationships.

The next chapter examines the benefits of being beautiful. Now that the key features of beauty have been presented, what are the benefits of being part of the attractive elite? Does being attractive make a difference? Is increased attractiveness worth altering oneself for? Why have people throughout history gone to such extremes to conform to beauty standards? As we will discover, the benefits of beauty are quite far reaching, and research reveals that the benefits range from fundamentally biological to a wide range of social perks.

2
The Benefits of Beauty

From childhood we are cautioned not to judge a book by its cover, likely because that is exactly what we are inclined to do. With experience, we learn that sometimes the cover can be accurate and informative about what we will find inside the book, but other times, it is radically misleading. This adage is frequently applied to our tendency to judge other people simply based on the way they look. Similar to an academic or fictional tome, such judgments can be accurate appraisals of the individual, but just as likely, they may be misrepresentations.

So, is there any value in judging others simply based on the way that they look? In simple terms, yes, we do learn a lot about others based on their physical traits and behaviors. Because these judgments can frequently be accurate, it becomes difficult to remember not to solely judge others based on their outward appearance. And, since we are all walking around judging, there are very clear benefits of possessing attractive traits. This chapter examines the benefits of being physically attractive, both from a biological perspective and from a social perspective. The characteristics that are considered attractive were presented in Chapter 1. Now, we explore why one may choose to enhance these traits for his or her own social benefit.

Very simplified, beauty tends to be correlated with biological health, youth, and reproductive potential. So, one of the benefits of being considered beautiful is that it likely means the individual is healthy, young, and fertile. Additionally, there are social perks to being attractive. Research shows that throughout the life span, attractive people are evaluated more positively by others, are treated better by others, are given more opportunities, engage in healthier and more positive behaviors, and have higher self-esteem. Each of these perks, and others, will be explored in this chapter.

BIOLOGICAL BENEFITS OF BEAUTY: AN INDICATOR OF HEALTH AND FERTILITY

Many of the biological indicators of attractiveness directly correlate with health and fertility. Symmetry, youth, waist-to-hip ratio (WHR), and body mass index (BMI) are all outward indicators of healthy development, developmental stability, and healthy genes. Symmetry, one of the predominant factors in attractiveness, has high correlations with health and developmental stability. Individuals with high facial and body symmetry tend to be physically and psychologically healthier and more stable. Regardless of whether the symmetry is a by-product of good genes or a healthy developmental environment, it tends to be a reliable indicator of health and stability.

Youth is also correlated with good health. Throughout young adulthood, the predominant causes of death are accidents and injuries rather than degenerative diseases or health-related concerns. Young adults tend to be physically healthy, do not tend to suffer from disease or deterioration, and tend to have high reproductive potential. Ailments and disorders tend to develop with advancing age, leading to more physical complaints and problems as youth (and thus, objective attractiveness) fades.

At a physiological level, lower WHR in women, typically rated as the most beautiful, is correlated with a decreased risk of heart disease, stroke, type II diabetes, gall bladder disease, kidney disease, breast and ovarian cancer, depression, and premature death. Women with lower WHR also tend to deal with stress better. Lower WHR in women also correlates with more regular menstrual cycles, increased levels of sex hormones, and more frequent and regular ovulation. Thus, although the hourglass figure is seen as attractive for a woman, the real value of having this ideal body shape is that it likely means the woman is a healthy, fit individual who is resistant to a range of physical problems. Overall, WHR is a relatively accurate indicator of hormone levels (past and present) and risk of current and future disease.

WHR also signals greater fertility. Women's WHR typically ranges from 0.67 to 0.80 and the ideal WHR as rated by males is around 0.70. A WHR of 0.70 creates the stereotypical hourglass feminine silhouette. Factors that increase or decrease a woman's WHR also impact her fertility level. Obesity, malnutrition, pregnancy, and the physical effects of menopause all directly influence WHR. For every 10 percent rise in WHR, there is a 30 percent decrease in the likelihood of pregnancy (regardless of age or overall weight). So a woman with a WHR of 0.80 (still within the norm for typical women) has a 30 percent less chance of becoming pregnant than a woman with a WHR of 0.70. Women with a more male-like physique or with fat around the waist with a WHR of 0.90 are 60 percent less likely to be able to successfully reproduce. WHR and shoulder-to-hip ratio (SHR) are influenced by sex hormones. High levels of sex hormones accentuate the female or male shape, also accentuating the likelihood of successful reproduction.

Attractive males also tend to have more reproductive success. Although earlier research found that attractive males do not necessarily have higher sperm quality, Pavol Prokop, from the University of Trnava in Slovakia, and colleagues found that attractive males are more likely to get married and more likely to have children. Attractive men are also more likely to have genetic diversity, disease resistance, and higher socioeconomic status. These are all traits that are more attractive to women and thus give attractive males more opportunities to reproduce, even if their sperm quality is not correlated with their outward appearances. Furthermore, attractive males are more likely to garner the attention of attractive females, who tend to be more fertile.

Finally, an individual's overall body mass (BMI) can be a strong indication of overall health. Being underweight or overweight is correlated with an increased risk for a whole host of disorders and diseases. Being classified as underweight or obese was correlated with a 51 percent increase in risk of death according to a study published by the American Medical Association. Being underweight can directly affect one's ability to successfully reproduce. The poor nutrition that tends to lead to an underweight body also contributes to delayed growth, a weak immune system, fragile bones, and anemia. Being underweight can also be a side effect of thyroid problems or other diseases. Obesity, alternatively, is often correlated with hypertension and heart disease, stroke, type II diabetes, different types of cancer, and gall bladder disease. Although obesity does not necessarily cause these disorders, its presence is a visible indicator that the individual may have unhealthy underlying genetics or lifestyle choices. Not surprisingly, low or high body mass decreases the ratings of physical attractiveness in psychological research.

SOCIAL BENEFITS OF BEAUTY: THE HALO EFFECT

Being attractive is correlated with many social benefits throughout the life span. An underlying concept that may explain the myriad of social benefits of being attractive is known as the halo effect. The halo effect, or the cognitive bias that people have in favor of beautiful people, tends to cause people to look more favorably upon those who are more physically attractive. Individuals' traits, personalities, and behaviors are interpreted to be more likable, humorous, and "good" if the individual is attractive. This creates a bias in favor of the more attractive person and makes him or her more likely to be trusted by others, more liked by others, and given more attention and opportunities by others.

Early psychological research by foundational psychologists such as Edward Thorndike discovered that attractive individuals were more likely to be rated as having strong leadership skills, higher intelligence, greater loyalty, and more dependability, regardless of their actual qualities. If someone is good-looking, it seems to be natural to assume that he or she is probably good in other areas as well. And, this effect can work both ways. If individuals are known to have good leadership skills or high intelligence, others tend to also rate them as more attractive. Having one visible positive quality leads others to perceive that one's other qualities are positive as well.

There are widespread implications of the halo effect in our society. For example, in schools, teachers are more likely to rate attractive students as hard working, intelligent, diligent, and engaged. Alternatively, attractive teachers are rated as more appealing and likable. In job interviews, more attractive applicants are more likely to be rated as competent and qualified for the position regardless of their actual skills or experience. In the workplace, attractive employees are likely to be rated as more effective, more enthusiastic, and more knowledgeable, and are given more opportunities and higher salaries. In marketing, attractive spokespeople are likely to increase the trust that a buyer places in the products and to increase positive evaluations of the products. In politics, more attractive candidates are overwhelmingly more likely to be elected. In a study of criminal cases, research even showed that jurors were less likely to believe that attractive individuals were guilty of a crime.

In the United States, physically attractive individuals are thought to be happier, smarter, and warmer. More attractive people are also perceived as healthier, as having higher self-esteem, and as being more sociable, less anxious, and less lonely. They are also thought to be more likely to interact with other attractive people, have access to more information from others, and be positively reinforced by others. They are less likely to be identified as psychopaths or as mentally unbalanced, and are thought to have better prognoses if suffering from physical or psychological problems. Attractive individuals are given more attention and are considered to be better adjusted. Although not all of these expectations actually hold up under controlled study, sometimes expectations are enough to create an advantage. For example, attractive people have not been found to be more intelligent than unattractive people, but attractive people do tend to achieve more. This is likely because achievement is not merely tied to intelligence. Achievement also requires opportunity, confidence, and persistence. Since attractive people are perceived to have superior intelligence, they are likely given more opportunities, more support, and more encouragement, all prerequisites for short- and long-term success.

The halo effect obviously has the potential to impact individuals in a variety of environments. Even after a person is aware of the natural tendency to perceive attractive people more favorably, it is still difficult to eliminate this bias. Like other cognitive biases, it is simply a function of making quick decisions while overwhelmed with a lot of incoming stimuli. Drawing conclusions quickly saves time and energy even if the resulting judgments are sometimes inaccurate. Seeing a beautiful person, a quality we consider to be good, biologically sound, and healthy, automatically causes us to bias our judgments in favor of the person in question. The symbol of a halo is associated with angels and goodness, so this tendency to see the good in others is what gives the halo effect its name.

Since ideal traits vary by culture, it is interesting to note that attractive people tend to be rated highly on those traits that are important to that culture. In individualistic societies, like the United States, attractiveness is correlated with higher ratings on traits of independence such as assertiveness, intelligence, extraversion, dominance, and likelihood of success. In less individualistic cultures, such as China, attractive people are rated higher on traits of interdependence such as more caring, loyal, trustworthy, and moral. These are traits that are more highly prized in those societies than individual gain and, so, are automatically attributed to more attractive people.

The only caveat to the benefits of the halo effect for attractive people is that attractiveness can lead to some negative stereotypic judgments as well. Some studies have found that, in tandem with the positive assessments discussed above, attractive people are also more likely to be judged as vain and manipulative. Other people, particularly those of the same gender, may also react unfavorably toward attractive individuals due to jealousy. Being overly concerned with one's appearance can be seen as being self-absorbed and some may be skeptical and cautious of being misled by an attractive exterior. Attractive people who appear to be unconcerned with their own appearance and open with others are rated the highest on all qualities. Unfortunately, attractive people may never know if others are interested in them due to their personality and inner qualities or just due to their physical appearance.

The following sections examine the implications of the halo effect across the life span. Research demonstrates benefits of beauty from childhood throughout many realms of adulthood. Attractiveness impacts teacher evaluations, treatment, behavior, and self-perception in classrooms, from elementary to college. The benefits of being attractive extend to employment opportunities, salaries, workplace assessments, and termination decisions. Attractiveness impacts formulation, maintenance, and termination of friendships and romantic relationships. Attractiveness also impacts the likelihood of interactions, cooperation, and expectations for future behavior by peers and colleagues. After looking at the different areas of research, this chapter will then examine whether attractiveness is actually related to these positive qualities or whether individuals are inaccurately being judged by the way they look.

BENEFITS OF BEAUTY IN CHILDHOOD

Judith H. Langlois and colleagues from the University of Texas at Austin undertook a large-scale meta-analysis and theoretical review to analyze a large portion of the research on the effects of attractiveness. Dr. Langlois and her team analyzed 919 research manuscripts ranging from 1932 to 1999. They focused on facial beauty and were interested to see if there were any overall effects of attractiveness on behavior, treatment, and personal characteristics. They examined both males and females and included studies of children and adults in their analysis.

In summary, Dr. Langlois found that attractiveness matters in several realms. First, she found that there were cross-cultural and within-cultural

Attractive children working together in an elementary school art class. Attractive children tend to be given more attention and more care by adults than their less attractive peers. They are more popular, more confident, and tend to be more successful in personal relationships and academic pursuits. (Monkey Business Images/Dreamstime)

agreements about what it means to be attractive. When asked to rate photos or other individuals, humans have general agreement about who is attractive and who is not, regardless of the race, culture, gender, or age of the rater. Second, the level of attractiveness has wide-reaching implications on social evaluations, treatment by others, behavior, and self-evaluations. Third, the effects were found for both children and adults, and for males and females.

Through interview and survey data, attractive children were found to be rated more favorably than unattractive children, regardless of how well the rater knew the child. Teachers rated photographs of unknown attractive and unattractive children as well as attractive and unattractive children whom they had in their classes. In both instances, the attractive children were rated more favorably with regard to behavior in the classroom, cognitive potential, and social potential. Attractive children were rated as having more social appeal; as having greater academic, developmental, and interpersonal competence; and as being significantly better adjusted.

The implications of these ratings became obvious when actual teachers and children were observed in the classroom. Teachers and peers treated the children who were rated as more attractive more favorably than they did the unattractive children. Attractive children had more positive interactions with others. Their social interactions displayed more positive emotions, the children and teachers seemed to enjoy each other more, and misbehavior was overlooked more frequently. Attractive children also had fewer negative interactions with others. This could be characterized by fewer displays of negative emotions, less arguing, and less aggression in their interactions with their teachers and peers. The attractive children were also given more attention and more care by adults. These children were called upon more frequently, given more time to ask and answer questions, and given more support in activities.

It is no surprise that this positive treatment was correlated with positive behavioral tendencies. When the behavior of the children was examined, attractive children showed more positive behaviors and had more positive characteristics than unattractive children. They were more likely to share, to include others, to cooperate, to be assertive, and to show positive emotions during their play. Based on these behaviors, they were rated as more popular, better adjusted, smarter, and more competent. Attractive children, being more popular with their peers, had more influence on and garnered more respect from others. Their opinions tended to be considered more readily, and others tended to be more agreeable with their ideas and suggestions. These positive assessments and expectations likely create a positive feedback loop. Receiving positive expectations and positive regard from others impacts self-esteem, mood, and behavior, while demonstrating high self-esteem, positive mood, and good behaviors elicits positive expectations and regard. This may be one explanation of why these children engaged in more positive actions. The positive expectations and positive treatment from others created a self-fulfilling prophecy.

Additional research added parents' judgments as well as judgments of peers to the mix and found similar results. Teachers, parents, and peers make judgments about and have expectations for children based on their level of attractiveness, which can directly impact children's expectations and judgments about themselves. Just like teachers, the impact of attractiveness influenced parents even for children they knew well. Level of attractiveness impacted the ratings even when the parents had personal experience with the children and had more information than just a photograph.

Furthermore, research shows that being aware of the bias created by attractiveness may not be enough to control its effects. Teachers frequently recognize that a child's physical features impact their assessments, but the features still play a role in shaping teachers' expectations. Margaret M. Clifford from the University of Iowa and Elaine Walster from the University of Wisconsin had elementary school teachers rate children for academic performance and social potential. Specifically, teachers were asked to approximate the child's IQ, their social status, the level of parental investment, and the child's future potential. The teachers were provided report cards and photographs for each child. Level of attractiveness was highly significant for all areas. The more attractive children were rated as having the most potential, having the highest IQ, being the most popular, and having the most highly invested parents, simply based on their physical features, even when additional (and even contradictory) information was provided.

In follow-up studies, similar results were found when children were asked to rate peers. The most attractive children tended to be rated as the friendliest, most helpful, and most independent. Unattractive children were rated as aggressive, noncompliant, afraid, and scary, and their popularity decreased as they aged. Unfortunately, children form their own self-conceptions based on how they are treated by others, so such ubiquitous treatment from family, teachers, and peers for unattractive children can damage developing self-esteem, potentially contributing to the decrease in positive behaviors, expectations, and self-efficacy as the child grows, creating a self-fulfilling prophecy. Unattractive children are not expected to attain as much, so they do not. On the flip side, attractive children similarly learn about themselves from others and the increase in positive behaviors and strong self-esteem could be a direct result of the same self-fulfilling social expectations.

As children exit primary and secondary school and enter a more adult world, one would expect that traits beyond physical features would be taken more into account. To examine whether or not this halo effect bias extended beyond the compulsory school years, researchers examined college students to see if the effects of beauty remained as significant.

BENEFITS OF BEAUTY IN COLLEGE

Considering that the halo effect is ubiquitous across the life span, it is unsurprising that the university environment is not immune. Level of attractiveness has a similar impact in university classrooms as in elementary school classrooms. Attractive students tend to be more confident, have higher self-esteem, ask more questions, and engage with their teachers and peers more frequently and with more positive emotions. They also tend to be more popular, have stronger social skills, and are more likely to be given and to take advantage of opportunities in academic and social environments.

The one distinction between the college and elementary classroom is the application process. Individuals must apply for entrance to the university. Comila Shahani from Hofstra University and colleagues from Rice University demonstrated that attractive applicants were more likely to receive positive evaluations for their entrance interviews than less attractive applicants. Even when interviewers had access to high school records, SAT scores, and GPA, attractive applicants were still rated as having higher qualifications than unattractive applicants regardless of their actual scores. This finding is particularly interesting, because, upon examination, attractiveness was not correlated with SAT scores, high school rank, or GPA. Thus interviewers were skewing their opinions based on attractiveness, regardless of the numerical data. The only exception to this trend was for highly achieving unattractive females. When high school rank was factored in for these high-achieving girls, they were ranked similarly to their attractive peers. Unfortunately for the university, during follow-up research, it was noted that attractiveness was not correlated with actual performance at the university level. Although attractive students were more likely to gain admittance to the school, the attractive first-year students did not perform better academically than the unattractive students. This demonstrates that attractiveness is not the best data to use as admittance criteria.

Although it may seem unfair that attractiveness factors in for the opportunity to gain higher education, the students themselves are guilty of the same bias. Nalini Ambady of Tufts University showed students videos of teachers and asked the students to rate them for effectiveness. The students, just like the interview panel, tended to base their assessments on attractiveness. Interestingly, however, the students' ratings were highly correlated with their end-of-semester ratings, 15 weeks later, showing that the first impression of others is highly influential and possibly quite accurate. This effect held true even when students had to make their ratings based on a six-second video of the teacher, leaving them primarily with only attractiveness cues on which to base their opinions. Dr. Ambady's research shows that within six seconds of meeting someone, individuals already make judgments and that initial impression is highly predictive of long-term assessments.

Although Dr. Shahani did not find correlations between attractiveness and academic performance for college students, there are other effects that likely factor in to overall college and occupational success, regardless of academic achievement. Dr. Anderson and colleagues from the University of California, Berkley, examined the long-term peer-status of individuals based on the level of attractiveness. Their research followed groups of individuals from fraternities, sororities, and dormitories to look at changes in level of respect, level of influence, and level of popularity over a nine-month period. Although attractiveness has repeatedly been demonstrated to enhance opportunities and status in the short term and between strangers, Dr. Anderson and his colleagues were interested in whether this effect held true over the long term as personality and personal characteristics become more familiar and prominent.

Dr. Anderson ultimately found that attractiveness was instrumental in attaining and maintaining status with peers, particularly for males. The highly attractive males tended to be the most popular and most respected, and had the most influence over their same-sex peers throughout the duration of the study. Personality traits also emerged as highly influential in the maintenance of status over the long term. Extraversion (e.g., highly outgoing and enthusiastic) was used by both genders to gain and maintain social status over the nine-month period. High levels of neuroticism (e.g., anxiousness and moodiness) were damaging to male's status, although high attractiveness buffered the effects.

Female attractiveness did not emerge as an important trait for a woman's social ranking among her same-sex peers. These differences are likely explained by the understanding that physical attractiveness tends to emerge as more important to a man's assessment of social partners. Men tend to place more emphasis on attractiveness when assessing male and female peers, while women may place less emphasis on physical traits and are more interested in the personality and personal characteristics of male and female partners.

Thus, the level of attractiveness is particularly important for male success during the university years, particularly with their peers. Attractive males are more likely to rank highly during the interview process, with attractiveness emerging as more influential than high school performance, SAT scores, or GPA. Attractive males are also likely to gain higher status within their peer group and to exercise more influence in their social circles. These opportunities allow them more experience to build leadership skills, gain assertiveness and confidence, and build connections for future cooperative efforts, regardless of their academic achievement. Females may not suffer the same effects, considering that high school performance and personality emerged as just as important as personal characteristics in gaining entrance to college and maintaining status among same-sex peers.

These differential experiences pave the way for future endeavors, and once students leave the college campus, they bring these skills into the workplace. They also bring their physical features, which will continue to wield an influence over their colleagues.

BENEFITS OF BEAUTY IN ADULTHOOD

In general, research studies demonstrate that attractiveness continues to be highly influential when individuals enter the workforce and begin building adult relationships. Attractive individuals have better job opportunities, earn higher incomes, and are promoted more frequently. They tend to be considered to be more intelligent and competent regardless of their actual skills. They are considered to be more socially adept, tend to have more friends, date more, and are more influential with their peers.

Based on Judith H. Langlois's research, the results for attractive adults mirrored those found for the children in her analysis. Langlois found that attractive adults were rated and treated more favorably than unattractive adults. When participants were asked to rate photographs of more and less attractive individuals for a variety of traits, the ratings of the attractive adults were much more positive. Attractive adults were rated as having greater occupational and interpersonal competence, having more social appeal, and being better adjusted than unattractive adults.

When Dr. Langlois examined the research on how attractive and unattractive adults were actually treated, she found that attractive individuals were treated significantly better than their unattractive counterparts. Others gave attractive adults more attention. Attractive adults were given more rewards and had more positive interactions, characterized by more positive emotions, more agreement, and more compromise. Attractive adults made more positive impressions, had fewer negative interactions, and were given more help and cooperation on problem-solving tasks.

Likely due to this differential treatment by others, attractive adults tended to engage in more positive behaviors. They were more likely to be assertive rather than aggressive and were more likely to engage others. They were also more likely to have occupational success, more friends, and more dating and sexual experience. Attractive adults also tended to be more extraverted; and had higher self-confidence, better social skills, and better mental health; and emerged as more intelligent than unattractive adults. Finally, attractive adults tended to have more positive self-evaluations than unattractive adults. When comparing self-ratings, attractive adults rated themselves as more competent and healthier than did unattractive adults. In general, ratings of attractiveness have a bigger impact on occupational success for men than for women. However, self-ratings of attractiveness have a greater effect for women than for men. This means that how a woman feels about herself affects her behavior and confidence. Feeling attractive eliminates some of the effects of her actual level of attractiveness. However, women also tend to make more accurate assessments of their own attractiveness, so women's self-assessments likely correlate with the actual levels of attractiveness as perceived by others.

Perception of attractiveness can, ultimately, be thought of as a positive feedback loop for behavior and intelligence. Those who are naturally attractive likely have the initial benefit of healthy genes and a stable developmental environment. If they also have access to good nutrition and good health care, take care of themselves, and are given opportunities, they are likely to maximize their attractiveness. The biological stability and environmental opportunities lead to higher self-esteem and more chances to learn and grown intellectually, leading to more intelligent individuals overall. These more intelligent individuals, in turn, take even better care of their minds and bodies and seek out wellness care, making them, overall, even healthier and more attractive. So attractiveness leads to opportunities and positive expectations, which allow one to maximize intelligence, which in turn leads to engaging in behaviors that improve overall health and attractiveness.

These benefits of beauty influence the attractive adult's overall general functioning and relationships. Next, specific benefits will be discussed. Attractiveness influences assessment, treatment, behavior, and self-esteem in a general way, but these differences translate into real-world benefits for those who are beautiful.

BENEFITS OF BEAUTY IN ROMANTIC RELATIONSHIPS

Since men tend to use physical traits more readily in their assessments of others, female physical beauty tends to be a significant asset in heterosexual romantic relationships. Michael Cunningham, from the University of Louisville, Kentucky, looked specifically at a woman's smile and eye size to make predictions about male behavior. He found that a wide smile and large eyes were linked to higher ratings of attractiveness and directly correlated with a man's estimation of the woman's intelligence, warmth, and social skills. Large eyes and a wide smile were also correlated with estimates of good health and fertility. Furthermore, larger eyes were directly related to a man's willingness to invest in a relationship. Men reported an increased willingness to take physical risks, provide for, self-sacrifice, go on a date or have sex with, and raise children with attractive women. Additional research demonstrated that having youthful features such as large eyes, high eyebrows, thick lips, round faces, and small chins elicits a protective response from others.

These youthful features are associated with honesty, innocence, warmth, and submissiveness.

For men, youthful-looking faces can have benefits as well. Baby faces with small noses and chins, big eyes, high eyebrows, and wide smiles elicit approach and nurturing from women. These open and more welcoming male faces tend to encourage interaction and approach from others. Other men and women rate them more highly on dimensions of honesty, kindness, warmth, and naiveté than they rated faces with more stereotypically masculine features. Evaluators also rated faces with these features as more lovable and as having more child-like attributes. However, while women may embrace their youthful features and use them to their advantage, baby-faced men are less likely to embrace these qualities, likely because they go against expectations of masculinity. In fact, baby-faced men are more likely to engage in delinquent behavior, and are more likely to be oppositional, hostile, and aggressive. Due to their more extreme behavior, they are also likely to earn more military awards and have higher academic achievement. Since baby-faced qualities tend to be more characteristic of the feminine ideal, women likely embrace these qualities and fulfill the stereotype, while men refute them and behave in a contrary way to avoid the perception of social or physical weaknesses.

In both genders, high attractiveness is likely to increase sexual interest from others. Not only are attractive individuals more likely to be more fertile, but also they are likely to actually have earlier and more frequent sex. Individuals rated as the most attractive, based on symmetry, WHR, and SHR, were found to have sex three to four years earlier than their less attractive peers. Furthermore, those same individuals were likely to have a greater number of overall sex partners and more frequent sex throughout their lives. Physically attractive individuals and individuals who perceive themselves as more physically attractive are more likely to engage in premarital sexual activity. This is likely a by-product of both increased confidence and increased opportunity.

Symmetrical males with a high SHR (broad shoulders and narrow hips) are more likely to be perceived as dominant and intimidating by other men. Their level of attractiveness makes them more appealing to women and more likely to incite jealousy from their male peers. They are likely to have more frequent access to sexual partners and to be shown more interest from others. Symmetrical women with low WHR (hourglass figure) are more likely to be perceived as attractive by males and likely to have the luxury of being more selective when choosing male partners.

A potentially unsurprising caveat to these positive expectations is that attractive men are also more likely to be less attentive to their partners and less faithful. Their level of attractiveness provides them with more options, so they are more likely to stray and to have sex with more partners. The same does not hold true for women. Attractive women tend to be initially more selective about their choice of partner, enforce high standards, and demand undivided attention. So attractive women tend to select higher quality partners who show signs of commitment and faithfulness, even at the expense of physical attractiveness, for their long-term romantic partners.

BENEFITS OF BEAUTY FOR COOPERATIVE AND COMPETITIVE OPPORTUNITIES

From childhood to adulthood, attractive individuals have access to greater opportunities and receive more cooperation from others. Beyond the positive social implications discussed above (friends, sex, relationships), having more opportunities also translates into accruing more experience and maximizing intelligence, which translates into getting better jobs and earning more money.

As having access to more opportunities and eliciting more cooperation from others is quite general and hard to measure, Matthew Mulford and colleagues from the London School of Economics and Political Science used a game paradigm to specifically assess the impact of attractiveness on how one is treated by others. The game they used was a version of the prisoner's dilemma. The prisoner's dilemma is traditionally used to assess the benefits of altruistic versus competitive behavior. The prisoner's dilemma pits strangers against one another in a competitive environment. It can be explained by imagining the participants as two accomplices who are being interrogated for a crime. They are not allowed to see or speak to one another before the interrogation. Since there is not enough evidence for a conviction, if both accomplices refuse to indict the other, they will both be set free. However, if only one person chooses to betray the accomplice, then the betrayer will go free and receive a reward and the other accomplice will receive the full punishment for the crime. If each player indicts the other, they split the punishment. So, the safest decision in this dilemma is to betray one's partner. That is the only way to receive a reward or shorter jail time. However, traditionally, when played over multiple trials, betraying one's partner tends to lead to one's partner retaliating, so over the long term, it is more mutually beneficial to cooperate with or not indict the other and receive no jail time at all. This strategy is risky unless one trusts one's partner because a betrayal will cause the player to lose the most.

To assess the impact of attractiveness on this process, Dr. Mulford invited participants to play this game. In his version of the dilemma, he allowed the participants to view photographs of their accomplice. After viewing the photograph, the participant was asked to choose whether or not to participate in the game with that accomplice. Rather than jail time, Dr. Mulford used money as the tangible reward. For example, if neither competitor betrayed the other, they each received a small reward. If they both betrayed one another, they received no reward. If only one of them betrayed, that person received the highest reward and the other lost money.

Two interesting things emerged from this study that demonstrated enhanced opportunities and enhanced cooperation for attractive individuals. First, individuals were more likely to choose to participate in the game after viewing a photograph of an attractive accomplice. Thus, from the very beginning of the study, it becomes apparent that attractive individuals are given an opportunity that less attractive individuals are less likely to receive. Being attractive was a preliminary requirement for securing a partner with whom to play the game.

Second, the level of attractiveness was predictive of the amount of cooperation. Attractive men were more likely to cooperate with their accomplice and they were most likely to cooperate when paired with an attractive partner. Attractive women were less likely to cooperate and only tended to cooperate with highly attractive partners, and so they tended to win almost double that of those women who rated themselves as unattractive. Since a high level of attractiveness made women less likely to cooperate and men more likely to cooperate, men tended to not earn as much as women in the game. Men who rated themselves as highly attractive were not likely to earn as much, demonstrating that, in this case, physical attractiveness may be more beneficial for women than for men in eliciting cooperation and investment.

The study concluded that attractive people have enhanced opportunities because people are not only more interested in engaging with them (more interested in playing the prisoner's dilemma game), but also, when the game is under way, people are more likely to cooperate if they are interacting with an attractive partner. The implications of this research can be easily applied to business. Attractive individuals are more likely to be approached and engaged and more likely to find agreeable and cooperative partners when employed in those opportunities than less attractive individuals.

Adults and children alike correlate attractiveness with trustworthiness. Judgments about trustworthiness are made very quickly. Within the first second of meeting or seeing someone, we already have a sense of whether we can trust them. Since these judgments are made so quickly, before we have even interacted with the other person or learned anything about them, they are largely unconscious and largely based on physical appearance. Even though the judgments are made quickly, there is high inter-rater reliability and high consistency across time. Furthermore, these initial judgments tend to be supported even after getting to know the person more. Since there is high reliability between raters, it is likely that there are key physical features that we all use to make these judgments.

Dr. Fengling Ma and Dr. Fen Xu from Zhejiang Sci-Tech University and Dr. Zianming Luo from the School of Environmental Science and Public Health examined which features lend themselves to these ratings of trustworthiness. They found that the features predominantly used for such judgments align with the same features used when rating the level of attractiveness. Specifically, people concentrate their attention on the eyes and mouth. They also tend to look at the cheekbones, brow ridge, chin, and nose. Furthermore, different races and all ages show consistency and all use these same cues when assessing trustworthiness and attractiveness.

Dr. Ma and colleagues created 400 East Asian faces with neutral expressions for participants to rate for the level of attractiveness and the level of trustworthiness, across two sessions. They found that faces with shallower chins, higher brows, longer foreheads, and longer noses were considered more trustworthy. For both children and adults, trustworthy ratings were correlated with lower ratings of aggressiveness and higher ratings of physical attractiveness. Thus, more attractive individuals garner a greater level of trust from others and reap the benefits in both cooperative and competitive situations.

BENEFITS OF BEAUTY ON OCCUPATIONAL SUCCESS

Psychology professor Megumi Hosoda and colleagues from San Jose State University, the University of Central Florida, the University of Albany, and the State University of New York examined research on the impact of attractiveness on occupational success and job-related outcomes. Dr. Hosoda's analysis of the research found that high attractiveness is always beneficial in job settings according to the last 40 years of research on the topic. He found that employers used level of attractiveness even when other information pertaining to skill level was available and that employers tended to reward the more attractive individuals with more responsibilities, opportunities, and promotions. Furthermore, both genders benefit from being attractive. Attractive males are considered more intelligent and attractive females are assessed as warmer. Overall, people who have above-average attractiveness are two to five times more likely to get hired than less attractive individuals.

Interestingly, Dr. Hosoda and his colleagues found that, although attractive individuals are assessed more positively with respect to job accomplishments, this trend has been decreasing over time. Job-relevant skills have started to take a more instrumental role in assessment of performance than appearance than they have in the past. However, promotions and opportunities still seem to be affected by the level of attractiveness. And, since level of attractiveness can sometimes be the deciding factor when employers must choose among applicants of similar qualifications, some companies have moved toward minimizing physical bias by eliminating any reference to gender or physicality when assessing new employees. Only then can qualifications and experience be assessed with less bias, at least when initially assessing candidates.

When specifically looking at attractiveness in women, men reported an increased willingness to recommend hiring and a greater willingness to invest money in an attractive woman versus an unattractive woman. Attractiveness in women tends to be associated with youthful features that suggest innocence and warmth. So, although attractive women may be hired more readily, they may be perceived as lacking in assertiveness and competence that may be needed to get the job done. Therefore, many women, once achieving a position of power, find they need to assume a more masculine style to enhance success. This may come in the form of feminine versions of masculine clothing, a masculine attitude, or a harder, decisive edge when speaking or making decisions. Highly feminine women may face more obstacles in stereotypically male positions due to the inherent gender bias in the workforce. So, attractive youthfulness is a double-edged sword for women because they are more likely to be hired but are less likely to be considered for a promotion or for a position of responsibility or power. In this case, unattractive women actually had an easier time pursuing and being successful in masculine jobs. This is known as the "beauty is beastly" effect, meaning that attractive people may be negatively affected by stereotypes just like unattractive people.

Once in the workforce, attractiveness has similar impacts on the success of salary negotiations. Upon hiring, attractive men tend to receive higher starting salary offers than unattractive men. Attractive men also tend to be given more responsibility and to receive better job evaluations, leading to accelerated promotion and salary increases. Attractive women were not found to receive higher starting salaries than unattractive women, but attractive women tended to be evaluated more positively, and thus were promoted more quickly and were shown to earn higher salaries than unattractive women within a few years on the job. Interestingly, these differences in salary amounts are not a by-product of increased self-advocacy or negotiations. Attractive individuals (and men, in general) are offered 12–16 percent more at their jobs without even asking for more compensation and they are also expected to be more competent and to accomplish more.

Attractiveness was found to have a definite effect on hiring decisions, salary, and promotion. Researchers next turned their attention to whether or not it has an effect on employment termination. Melissa Commisso and colleagues from Rock Valley College and Northern Illinois University designed a study to see if level of attractiveness had an impact when making termination decisions. To simulate decisions made by employers, Dr. Commisso asked college students to make termination decisions based on fabricated employee files that contained poor performance reviews. Each file also contained a picture of the employee and the picture was an extremely attractive individual, an average individual, or an unattractive individual. In line with other research, undergraduate students were more comfortable and more likely to suggest termination of the unattractive employees based on the poor performance reviews. They were not as likely to suggest termination for the average or highly attractive individuals, confirming, again, that attractiveness leads others to judge less harshly and to provide more opportunities.

In general, errors and poor performance are more detrimental to unattractive individuals. As predicted by the fundamental attribution bias, or our tendency to blame a person for faults rather than to blame the situation, unattractive individuals are likely to be blamed for errors, and poor performance is taken as a reflection of their personal traits. Poor performance for an attractive person, however, is likely to be explained away, and the blame is placed on the situation rather than on the individual's personal qualities. Typically, the environment is more influential on behavior than it is typically given credit for, but unattractive people are not given the benefit of doubt as frequently as attractive individuals.

OTHER BENEFITS OF BEAUTY

The occupational benefits that stem from being attractive are only the beginning. Research has demonstrated that the social benefits for attractive people stem far beyond the workplace. For example, attractive adults are four times less likely to be suspected or punished for crimes, and receive less harsh punishments than unattractive people if they are indicted. Likely as a by-product of the halo effect (the bias that attractive individuals must have other attractive qualities as well), it is difficult to believe that attractive people could have bad qualities, so they are more likely to be given a warning or to be excused. This is especially true for attractive women who look innocent and warm. They are less likely to be considered suspects in crimes and less likely to be blamed or convicted.

In a review of the existing literature regarding the attractiveness bias, Sean N. Talamas and colleagues from the University of St. Andrews in the United Kingdom echo the pervasive influence of attractiveness on our decisions in various realms. Attractiveness influences our opinions and decisions in voting for political candidates, electing others to leadership positions, making decisions about consequences for criminals in court cases, deciding on punishments for children who have misbehaved, establishing expectations for children, being promoted in the military, and evaluating academic performance and maturity level.

Overall, attractive people are more likely to receive more respect from peers, coworkers, and supervisors, and have happier relationships. They are perceived as more competent socially and intellectually, and have higher expectations from others. People are more willing to engage in personal or work relationships with attractive people, and attractive individuals tend to have happier, more stable personal and romantic relationships. Attractive people (both children and adults) are thought to be more competent and effective. Attractive males tend to be perceived as more cognitively competent and attractive females tend to be perceived as more socially competent. These effects are particularly strong where there is no other information to go on. Greater expectations are held for attractive people, and they tend to be assessed less harshly, favored, respected, and given more power.

While attractiveness has been demonstrated to be highly beneficial in many realms, there are some negative associations correlated with being attractive. Some may judge attractive people as being more egocentric and self-obsessed and they are, in fact, more likely to rely on their looks. They may also be judged to be less honest and less concerned about others, particularly if they are seen to make efforts to maintain their appearance. Attractive people are also likely to experience more jealousy from those around them and may be treated negatively as a result. Attractive people also may not know whether others like them for their personal qualities or just because of their appearance.

IS THERE TRUTH IN BEAUTY?

There is no doubt that attractive individuals receive preferential treatment, on average. As demonstrated by Dr. Langlois, even facial attractiveness has significant impacts on attitudes, treatment, behavior, and self-perceptions for both adults and children, regardless of gender. Attractiveness is related to perceptions of popularity, school and occupational success, intelligence, adjustment, and experiences, regardless of how well the raters knew the individuals they were rating. The effects of attractiveness on ratings, treatment, and behavior are just as strong whether or not the individuals know each other. Findings were also consistent for males and for females, suggesting that male attractiveness is just as important as female attractiveness for social interactions, mate selection, and assessment of overall health.

However, these perceptions of competence, adjustment, effectiveness, and health based on attractiveness level do not necessarily correlate to actual competence. In reality, only those in the lower half of the attractiveness scale had correlations between attractiveness and health and intelligence. Individuals who range from unattractive to average do tend to be less healthy, have more developmental complications, and have lower intelligence scores. So, facial attractiveness may serve as a valid signal for intelligence level within this lower extreme. This correlation does not hold for average to highly attractive individuals. So, our assumptions that intelligence and health increase with attractiveness are overgeneralizations and not accurate reflections of reality despite their ubiquitous presence.

The perceptions of competence, adjustment, effectiveness, and health for both attractive and unattractive individuals may partially be the by-product of a self-fulfilling bias. Since attractive people are treated with respect, have high-expectations, are given opportunities, and are provided attention and care, they likely develop high self-esteem and maximize their potential. Unattractive people are treated more poorly, are more likely to be abused as children, are more likely to be bullied, receive lower expectations from others, are given fewer opportunities, and are given less attention and care, and thus, they likely have lower self-esteem, lower motivation, and lower achievement overall. The accomplishments of both groups are an interaction of genetic potential and environmental influence and different experiences could lead to different outcomes. Since higher expectations have been shown to increase future performance, attractive individuals tend to stay in school longer, attain higher degrees, are awarded with more opportunities, and have a better chance of succeeding academically and occupationally.

CONCLUSION

Psychological research demonstrates that there are physical, social, and psychological benefits of being attractive. Physically, traits that are typically considered beautiful are indicators of overall genetic health and a history of developmental stability. Traits typically associated with attractiveness, such as WHR and BMI, also tend to be correlated with reproductive health and future reproductive potential.

Socially, people who are considered attractive tend to prompt expectations of the halo effect from those around them. Since beauty is associated with goodness, beautiful people are assessed more favorably in other aspects as well. Beautiful people are presented with more opportunities, are more kindly treated, are more favorably assessed, and are more popular with peers. They tend to be offered higher salaries, have higher academic achievement, and are rewarded with more promotions and raises. Attractive people tend to elicit more interest from others—professionally, socially, and romantically—have more sexual partners, and have more children overall.

Psychologically, attractive individuals tend to have higher self-esteem, a stronger sense of self-efficacy, and overall better mental health than less attractive individuals. Attractive individuals have more confidence, rate themselves as more competent, and have more positive self-evaluations. These physical, social, and psychological benefits likely create a positive feedback loop where physical benefits have a positive influence on the quantity and quality of social interactions, which positively impacts psychological health and stability and thus further contributes to physical developmental stability and increased social opportunities. Physical attractiveness serves to increase confidence, which invites increased interactions from others, contributes to better social skills, and invites more positive interactions. So, physical traits contribute to physical attractiveness, psychological health, and social skills.

3
Buying Beauty

Since physical attractiveness contributes to a myriad of physical, social, and psychological benefits, improving and manipulating physical traits to increase our own levels of attractiveness may be highly beneficial for overall success. While many factors of attractiveness, such as social status, body shape, and skin color, are difficult to change, other features, like clothing, behavior, complexion, and facial features, can be easily manipulated, highlighted, and enhanced to increase one's own perceived level of beauty. In fact, individuals have been engaging in beauty-enhancing rituals throughout ancient and modern times and reaping the physical, psychological, and social benefits. This chapter includes the methods of increasing one's own attractiveness to others, both currently and historically. In it, we will explore typical methods as well as more extreme practices.

Beauty is correlated again and again with those traits that make an individual appear young, healthy, and fertile. For those on a quest to be beautiful, mimicking those traits can successfully fool the human eye. There are cross-cultural similarities in all image-conscious societies to preserve youth and enhance attractiveness. There are differences as well, however. Women from the United States are more likely to report feeling and looking stressed and tired than European women. They are also more likely to start using beauty products earlier and to attempt to prevent aging rather than to correct or restore their youthful appearance later. Women from the United States are also more likely to switch products if there are not immediate results. They are more likely to worry about their skin, proactively fight aging, and have drops in self-esteem with the onset of wrinkles and creases than are European women. As a result, Botox injections are at the forefront of aesthetic treatments in the United States.

This chapter will examine the methods of increasing one's own level of attractiveness and present research on the effectiveness of such behaviors. Here, we examine a host of traits that are considered most beautiful as well as how individuals attempt to achieve them on a personal level.

CURRENT METHODS

Natural Approaches

There are many natural methods to enhancing one's own physical attractiveness. Some natural methods include getting adequate amounts of sleep, engaging in physical exercise, monitoring one's diet, controlling one's own facial expressions, and maintaining an upright posture. Sleep enhances natural beauty, particularly in the face. Individuals who were rated after a good night's sleep were rated as more attractive than after a period of sleep deprivation. Sleep allows the body to physically restore itself and promotes health and overall natural attractiveness for men and women.

In addition to sleep, exercise is a potentially effective way to increase attractiveness. Exercise increases muscle tone and accentuates gender differences in body type. Toned feminine women and muscular masculine men are rated as more attractive than heavier, less defined individuals of either sex. Exercise targeted at particular body parts can help accentuate body shape. By toning abdominal muscles and building gluteus muscles, women can create a more flattering waist-to-hip ratio, shaping and improving on their actual underlying genetics. Men can narrow their abdominals and create upper body muscular bulk to create the attractive inverted triangular torso indicative of a high shoulder-to-hip ratio. Regular exercise also improves lung functioning and increases heart health, which gives the skin a healthy, ruddy, well-oxygenated glow. Exercise can also be used to achieve a healthy weight, achieving an attractive body mass index.

Along with exercise, a healthy, natural diet can help maintain a healthy body weight and provide the vitamins and nutrients needed for strong healthy skin, eyes, muscles, and hair. Drinking plenty of water tends to enhance skin suppleness and delay the appearance of wrinkles. Drinking water and eating a balanced diet also contributes to a healthy complexion and increased energy, both of which have been identified as having a positive impact on ratings of attractiveness.

Facial expressions can also transform the level of beauty of the face. A smiling face is typically rated as more attractive than a neutral or frowning face, so a smile can enhance natural beauty. In a psychological study, women who smiled in a social setting were more likely to be approached and were rated more favorably than women who did not. Happy faces are perceived as being more dominant and engaging and tended to invite confident, friendly, and stable interactions from others. Raised eyebrows, wide eyes with crinkled corners, and upturned lips are all components of a smile. While a genuine display garners the highest ratings of attractiveness, intentionally displaying these facial characteristics improves the ratings of attractiveness as well. A pleasant face also appears to be warmer, invites interaction, and garners more attention and more approach behaviors from others. In our social world, these characteristics can create more opportunities in work, play, and intimate relationships regardless of the underlying level of attractiveness.

Posture, similarly, contributes to attractiveness. A healthy posture consists of keeping one's chin parallel to the floor, while keeping the shoulders back and aligned. Good posture generates maximum natural height, making an individual appear taller and healthier. Good posture also tends to make an individual appear more dominant, confident, and approachable. An open posture also contributes to attractiveness. Keeping one's torso open to the world demonstrates availability and invites others in, something that most individuals rate as attractive. Crossed arms or individuals who are bent low with their attention on their feet (or their cell phones) detract from the ratings of attractiveness. Exposing the neck with a lifted chin also makes one appear more open and makes one even more attractive to others.

Not only can posture attract others, it can also have an effect on one's own mood and confidence level. Intentionally standing in a straight and open posture (such as with hands on hips) improves mood and increases confidence. So, posture can both communicate what one is feeling and also influence one's emotions. Research has shown that we tend to mimic the posture of a conversation partner, particularly when we like, respect, or are attracted to that individual. Furthermore, we tend to be attracted to those who mimic us. Dr. Marco Bertamini, Dr. Christopher Byrne, and Dr. Kate M. Bennett from the University of Liverpool even found that we tend to rate those who are in the same body position as we are as more attractive. Even sitting or standing can significantly affect our ratings of a partner if we ourselves are sitting or standing.

To study the effects of posture on the ratings of attractiveness, Dr. Don R. Osborn from Bellarmine College in Louisville, Kentucky, had undergraduates rate photos for level of attractiveness as well as level of sexiness. These photos varied in posture and included slouched, normal, and military postured individuals. Results confirmed that a slouched posture communicates negative attributes about the individual in question. In this study, slouched individuals were rated as more likely to have health problems, more modest, less intelligent, less socially skilled, less assertive, less fertile, less likely to have an affair, and, finally, less attractive. Thus, this study shows that a behavioral factor such as posture can influence the level of attractiveness and may be just as important as body shape. Furthermore, the models used for this study varied in overall weight, and no difference was found for body size in attractiveness ratings. So shape and posture emerged as more important for level of attractiveness. Normal and military postured individuals were rated as more attractive, sexier, and garnered more attention from respondents.

Cosmetics

Natural methods can only enhance one's natural appearance. For those who want to make more substantial changes or want to alter their natural look, there are many other popular methods that have been developed to increase one's perceived level of physical attractiveness. First of all, a whole host of cosmetics can preserve one's appearance and create many of the features that are markers of natural beauty. Since men tend to rely more on physical features to assess attractiveness than women, it is no surprise that the cosmetics industry is focused primarily on women. Lotions to de-age and provide supple, youthful skin are marketed in a variety of forms and under a multitude of labels. Lotions focused on eliminating dark spots and smoothing wrinkles are widely used to delay the physical signs of aging and to preserve or reclaim the appearance of youth and beauty.

In addition to lotions that attempt to preserve the quality and health of the skin, cosmetics are also a very simple way to subtly alter one's appearance to more closely mirror an ideal. Imperfections can be smoothed over, extremes can be softened, features can be enhanced, and symmetry can be created. Foundation can create a soft, even, skin tone, void of the imperfections that may decrease physical appeal. Thin lips can be enhanced by rich, bright color, creating the universally attractive full lip. Cheekbones can be highlighted and raised, giving the face a sexually mature, yet youthful glowing, vivacity. Eyebrows can be plucked and sculpted to create an open and inviting expression or darkened to provide a more youthful appearance. Eyes can be outlined and highlighted to produce the perception of larger eyes, which contributes to a more youthful look. Even without conscious intention, women across the United States and other parts of the world attempt to conform to the biological ideals of beauty during their morning makeup routines.

A simple Google search reveals that the use of cosmetics is not a process to be taken lightly. Varying sites claim that they have the easy 7 to 14 steps to prepare one's skin before even applying makeup. One is instructed to wash, exfoliate, tone, moisturize, and prime before cosmetics are even applied. Typical cosmetics include lotions, foundations, concealers, highlighters, eye shadows, eye liners, mascaras, blush, lip liners, lipsticks, and lip glosses. Then, once everything is in place, one should not forget to seal. This routine may take time, but the process promises a face that pops, is contoured, is sculpted, is blended, shimmers, and smokes, or, if desired, special steps can be taken to create a "natural" look so one can complete the 14-step process and appear to not be wearing any makeup at all. When one is done being striking, however, one must then begin the process of makeup removal. This also incorporates a "simple" 7- to 14-step process to thoroughly remove the cosmetics to maintain a healthy youthful complexion. Many women are so inundated in this process that they are uncomfortable being seen in public without undergoing the morning ritual of "putting on their face."

Due to the widespread and daily use of cosmetics, the beauty industry made over $46.2 billion in 2015 and is projected to make $51.8 billion in 2020. Cosmetic purchases alone rose by 13 percent in 2015. Instagram and YouTube are credited for driving the trends of experimentation and changing the use of highlighters, concealers, sculptors, primers, and more. Consumers buy the products regularly and experiment with different colors and brands. In the current trend, there is also more emphasis on immediate results rather than just long-term care. If lotions do not provide immediate results, women are likely to change brands quickly.

Since immediate results are in demand, fragrances are sometimes rated as more effective than makeup. There is a current trend for increased perfume and cologne use because the effects of these products are immediate and changeable. To compensate for the consumer shift, many cosmetic companies have started marking their own brands of perfume and cologne. There is also a current trend for natural products rather than chemically created products, again creating a shift in the cosmetics industry. To be successful, cosmetic companies need to offer scents, natural products, luxurious (more expensive) items, and items that are more affordable. Based on consumer interviews, the typical consumer has an assortment of products and tends to purchase inexpensive items for experimentation and more expensive items for long-term use. This means that, to be a competitive brand, manufacturers need both fragrances and cosmetics, expensive and affordable options, as well as multicultural shades.

Do Cosmetics Work?

Cosmetic use can be a complicated, expensive, and time-intensive endeavor, and research has focused on whether using cosmetics actually provides a benefit. Laboratory studies examining ratings of photographs of women with and without makeup have consistently found that women wearing makeup are considered to be more attractive, more sexually desirable,

A woman applying eye shadow and other cosmetics can increase her perceived level of attractiveness, intelligence, and aplomb. She risks, however, appearing conceited and less faithful. (Phovoir/Dreamstime)

and more likely to be approached in social settings. Laboratory research specifically demonstrated that women with full makeup were rated as more attractive than those who were just wearing eye makeup, that those wearing eye makeup were rated as more attractive than those just wearing foundation, and that those wearing foundation were rated as more attractive than the same face without makeup. Interestingly, the difference in ratings is primarily contingent on the sex of the participant. Specifically, men are more likely to rate a photo of a woman wearing cosmetics as more attractive than when she was presented without makeup. Women's ratings of attractiveness were not contingent on the presence or absence of cosmetics.

Dr. Don Osborn from Bellarmine College in Kentucky also examined the use of makeup on the ratings of attractiveness. He had undergraduate students rate photographs for level of attractiveness as well. Amidst distractor photos, he included photos from a before/after beauty campaign. Some subjects saw the *before* photos and some saw the *after* photos. These photos were especially suited to this design because other elements of the individuals' appearances were not changed. Their hairstyles, clothing, expressions, and features were identical in both photographs. Prior studies had found that men are more influenced by makeup than women. As in this previous research,

Buying Beauty

in Osborn's study, male ratings of the photographs were more extreme than female ratings. Males also had more extreme ratings of the photographs of women with makeup. Both genders rated the photos of women with makeup as smarter, more socially skilled, sexier, and overall more attractive, but men's ratings were higher in all of these areas. Both genders also rated the photographs of women with makeup as significantly less modest than those with no makeup, but on this dimension, women tended to give the more extreme scores. The women in the photographs with makeup were also rated as being more assertive and having fewer medical problems, but also as being more likely to have extramarital affairs.

In Dr. Osborn's study, he acknowledged that the use of cosmetics might have a behavioral influence on the women who use them. By using makeup, women can actively influence how others perceive them, and this certainly has an influence on how they feel about themselves and how they may behave or hold themselves. When people know that others perceive them as more attractive, this can influence the way they feel and act in social situations. Feeling beautiful has been shown to increase a woman's level of confidence and subtly alters her social interactions. Feeling and acting confident garners more attention from others, invites more approach behaviors from others, and increases ratings of satisfaction in interactions. So, cosmetics likely increase the perception of physical attractiveness while also increasing the attractiveness of the individual's social behavior. Both likely help explain the effectiveness of cosmetics use.

To take into account behavior and to more thoroughly understand the effects of makeup use beyond mere photographs, researchers have taken this body of research a step further and have designed real-word studies that look at behavior outside of the lab. Although participants say they prefer women with makeup and report they would be more likely to engage with them, researchers were interested in actual behavior outside of a laboratory setting.

To this end, Nicolas Guéguen, a French professor of social and cognitive psychology at the University of Southern Brittany, evaluated behavior at a bar. He asked female colleagues to sit at a bar, where he carefully controlled for differences in their behavior and recorded the number of men who approached them over the course of one hour. He found that when wearing makeup, women were approached more often and more quickly than they were on nights when they did not wear makeup. His research shows that if one's goal is to get attention, attract a potential mate, or get noticed, cosmetics can be extremely effective.

Furthermore, Céline Jacob and colleagues, also from the University of Southern Brittany as well as from the University of Maine in France, designed another field experiment in a restaurant setting. They were also interested in the effects of cosmetic use on behavior. Specifically, they looked at tipping behavior to examine whether a server's use of cosmetics would have an impact on the patron's behavior. Previous research has shown that tipping is influenced by breast size, hair color (blondes may or may not have more fun but they do get tipped better), small body size, and overall level of attractiveness, at least for women. Tipping behavior is also influenced by facial expression, and body or hair ornamentation, particularly from male patrons. In her study, Dr. Jacob assessed two waitresses over a period of four weeks. During this time, the waitresses either went makeup free or a beautician was hired to apply eye, cheek, and lip makeup to accentuate their attractiveness. The experiment was carried out in France where additional tipping is not common since the gratuity is already built into the menu price. However, a significant difference was found for male patrons in this experiment. The men in the study tipped more often and gave larger amounts of money to the waitress when she was wearing makeup than they did when she was not wearing makeup. Consistent with earlier studies, however, female patrons were not influenced by the presence or absence of cosmetics and there were no significant differences in their tipping behavior. So, female makeup use seems to specifically target males. Whether it simply draws attention, inspires a need to provide, or changes the way the female interacts, cosmetic use leads men to tip more and approach more quickly.

As a final example of research examining the effect of cosmetic use, the impact of applying makeup was demonstrated by asking college students to help make decisions about candidates in a simulated interview setting. Students were provided with information, including photographs, of candidates for a job. This research found that women wearing makeup were more likely to be recommended for the position and were offered 8–20 percent more starting pay than those not wearing makeup, when all other factors were consistent. This study was conducted using photographs, so behavior of the candidate was not a contributing factor. Cosmetics alone were responsible for the difference in suggested pay.

This research demonstrates that cosmetic use provides individuals with the power to influence the behavior of others by manipulating their own levels of attractiveness. Also, not only does makeup alter the behavior of others, as demonstrated in the research above, but it also alters the confidence levels and behaviors of the individuals wearing it. This increase in physical appeal boosts confidence, which changes the way one interacts with others, and consistently improves ratings of attractiveness and alters the behavior of others, particularly males. Viewed from this perspective, it is not surprising that the cosmetics industry is a booming business around the world. For women, in particular, using cosmetics can improve others' perceptions of their levels of attractiveness, assertiveness, health, social skills, competence, and sexuality. Thus, makeup can be a relatively inexpensive mechanism for radically altering opinions and the behavior of others. In addition to clothing choices, which will be discussed in the next chapter, and posture, makeup can profoundly enhance signals of beauty and increase personal confidence levels.

The use of makeup leads to an interesting philosophical debate. As we enhance our physical features to be more pleasing to others, we are simultaneously misleading those around us. Changing our outward appearance does not change the underlying genes, and, therefore, we are tricking others into passing our less-than-perfect genes into the next generation. This simple tactic, while beneficial for the individual, is essentially altering our evolutionary path by increasing reproductive success for less attractive individuals, who can now successfully reproduce and pass their less healthy genes on to the next generation. Thus, at an individual level, makeup use can enhance one's own success, but at a species level, we may be subtly and detrimentally altering our evolutionary course.

How Do Cosmetics Work?

Cosmetics seem to work by exaggerating the level of sexual dimorphism (i.e., heightening typical gender differences) and accentuating the appearance of youth and femininity in the female face. Women are characteristically fairer and softer than men, and makeup can artificially enhance that difference. Makeup can create a youthful look by smoothing the skin, obscuring wrinkles, and hiding scars or discolorations for women who either do not naturally possess these characteristics or are further along in the aging process. Even makeup routines that achieve a "natural" makeup-free appearance still enhance one's appearance by creating a smooth, youthful, healthy, and glowing face.

Alex Jones and colleagues from Gettysburg College statistically demonstrated that makeup could be used to increase sexual dimorphism in the face. Their research demonstrates that cosmetics can create the appearance of luminance of the skin and greater contrasts between the eyes, lips, and cheekbones. These traits are characteristic of enhanced femininity. Dr. Jones also revealed that eyebrow grooming further accentuates sexual differences by lightening the distinction between the brows and the face. Since men tend to have thicker, bushier eyebrows, decreasing the thickness of the eyebrow can create a more feminine look, especially for younger individuals. With age, eyebrows tend to naturally thin, so older women may actually need to enhance their brows to maintain a more youthful look. Ideal beauty is a tricky balancing act between increasing sexual dimorphism and maintaining a youthful appearance. Eyebrow plucking may actually increase the maturity of a woman's face, but it also increases her level of perceived femininity.

Young women can trade age for femininity, but older women are rated as more attractive if they choose the more youthful, thicker-browed look.

To enhance the perception of attractiveness, one could expect that cosmetic use should be most effective when it is unnoticeable. Makeup that enhances the appearance in a way that is undetectable should be most effective because viewers would think that the appearance is natural and indicative of underlying genetic quality. However, this does not seem to be the case. Although subtle makeup does significantly enhance the ratings of attractiveness, even blatantly apparent cosmetic use provides similar results, even when the makeup is clearly not a by-product of genetic quality. In this case, makeup use may also be a symbol of status, grooming habits, and conformity to biological and social norms, all of which are attractive to others.

Cosmetic Surgery

While cosmetic use can be expensive and time consuming, it is a relatively noninvasive method of altering one's level of facial attractiveness. For many, such simple measures are not enough to achieve the goals set forth in society. More and more individuals are deciding to engage in cosmetic surgery to permanently alter facial or body features to more closely resemble a cultural ideal. Cosmetic surgery includes surgical and nonsurgical treatments that alter appearance for aesthetic (rather than functional) reasons. These elective procedures typically have the goal of improving one's appearance or increasing one's confidence and differ from reconstructive procedures, which have the goal of restoring function or reconstructing body parts after damage.

Cosmetic surgery is certainly a more invasive and more expensive method than cosmetic use to achieve a physical ideal. Given its invasiveness, a logical individual would expect it to be rare and used only in cases of deformity or dire necessity. However, according to the American Society of Plastic and Reconstructive Surgeons, over 17 million cosmetic procedures were performed in 2016 in the United States alone, a 3 percent increase from 2015. There are a multitude of surgical interventions just for the face, including rhinoplasty (reshaping the nose), otoplasty (reshaping the ears), blepharoplasty (reshaping the eyes), mentoplasty (reshaping the chin), and rhytidectomy (tightening the skin on the face, face lifts). The top five elective procedures in 2016 were breast augmentation, liposuction, nose reshaping, eyelid surgery, and face lifts. Breast augmentation fees average just under $4,000 and the industry made well over $1 billion for that procedure alone in one year. Nose reshaping costs around $5,000, bringing in over $1.125 billion. Surgery fees for tummy tucks cost an average $5,750, totaling over $740 million in 2016. The procedures with the biggest increases in number over the last year include butt augmentations, which were up by 26 percent, lower body lifts, up by 34 percent, and labiaplasty (reshaping of the labia), which were up by 39 percent. Of the total procedures, 92 percent were for women and only 8 percent were for men.

Data from the British Association of Aesthetic Plastic Surgeons show that Americans are not alone in engaging in elective surgery to enhance physical appeal. In 2015, 51,000 British individuals underwent cosmetic surgery, a 13 percent increase from 2014. Women accounted for 12.5 percent of that increase and breast augmentation remains the most popular procedure for women in the United Kingdom. Other areas of increase were face and neck lifts and liposuction. Men are also seeking more brow lifts, eyelid surgeries, nose reshaping, liposuction, and male-breast reductions.

In China, the dramatic increase in cosmetic procedures has led to a new term called *renzao meinü*, meaning "artificial beauty." This term refers to women in China who engage in extensive cosmetic surgery to meet cross-cultural beauty standards. The most common procedures in China, as well as in countries like Korea and Japan, are facial reshaping, eye enlargement, eye-lid surgery to create the appearance of a Caucasian double lid, and nasal reconstruction. These women report that they want big eyes and a more "Western" look. Women have spent billions of dollars changing innate, healthy physical features. When asked, women in China and Korea reported that they sought out surgical and nonsurgical procedures for a variety of other reasons as well. They reported a desire to look younger, to conform to feminine cultural ideals, to take control of their own bodies, and to increase social status.

Given the impact that facial features and body shape have on the ratings of level of attractiveness, it is not surprising that the top procedures are those that transform the overall body shape (enlarging the breast or removing fat from around the waist) and enhance facial features. Cosmetic procedures alter physical attractiveness and likely also impact psychological functioning, body image, and self-esteem.

Although these procedures may sound ludicrous to a healthy, attractive individual, research shows that individuals who undergo cosmetic surgery, especially those who radically improve their level of attractiveness, do tend to receive long-term benefits. Those who substantially increase their level of attractiveness tend to be more likely to be hired for a job, are offered more money, receive more raises, and attract spouses who earn more money. Neelam Vashi from Boston University reports that these individuals typically recover the surgical costs within a couple of years. Those who do not have such a radical change from surgery are not typically able to recoup their costs. However, reported findings do show that people who seek cosmetic surgery and who are happy with the result tend to date more, have increased self-esteem, and have increased life satisfaction and happiness.

Depending on how cosmetic surgery is defined, it can refer to traditional surgical procedures to those of a more nonsurgical nature. The overall number of surgical procedures is actually down by about 10 percent from 2000, but minimally invasive, nonsurgical procedures, such as chemical peels, laser hair removal, and microdermabrasion, were up by 158 percent in that same time span. Botox injections, soft-tissue fillers, chemical peels, microdermabrasion, and laser treatments make up over ten million procedures annually in the United States. These treatments are used to decrease the appearance of wrinkles, add fullness to lips, enhance the contours of the face, improve the texture and color of the skin, minimize scarring, and smooth out pigmentation. Averaging the number of all cosmetic procedures, surgical and nonsurgical, the rates of cosmetic procedures are up by 115 percent over the last 15 years.

Not surprisingly, across all cultures and ethnicities, procedures tend to be directed at efforts to enhance youthful features and decrease signs of aging. Since youth is a particularly important feature used by men to rate a woman's level of attractiveness, it makes sense that the majority of patients are women. And, given the predominance of Caucasian women in the modeling industry, many coveted features are those typical of the Caucasian race. Expanding the modeling and advertisement industries to include more representations of beautiful women from more races might result in a decrease in the number of procedures sought by women cross-culturally.

Maintaining and Removing Hair

Hair is unarguably an important aesthetic feature for humans, particularly women. Shiny, healthy, strong hair naturally advertises underlying physical health and strong genes. Limp, greasy, breaking strands give natural cues to possible disease or poor grooming habits. Hair color, placement, removal, and quality all impact level of perceived attractiveness, although preferences vary around the world. Many individuals dye their hair regularly to maintain a healthy, youthful appearance. Typically, youth is associated with lighter, longer hair. As women (and men) age, hair tends to naturally darken, signaling maturity. Since youth is associated with higher reproductive potential for females, lighter hair tends to be rated as more attractive overall for women. Thus, generations of women have used products to highlight, lighten, and enhance their natural locks. Youth is also associated with longer tresses, particularly for women. Longer hair has been found to communicate sexual maturity, intelligence, health, and vitality. Hormones control hair growth, so hair health is a clear indicator of underlying physiological functioning. With age, many women cut their hair shorter for a variety of reasons and the result is that they are seen as more masculine, older, less sexually interested, and less fertile. Women pay to have their hair cut, colored, and styled and purchase products to nourish, strengthen, and create a shiny sheen. Hair gels, sprays, and mousses provide body and bounce to naturally limp strands.

Beyond caring for the hair on their heads, many women in the United States regularly remove hair from other parts of their bodies to increase the appearance of youth and of overall attractiveness. Women commonly shave armpits and legs to maintain unnaturally smooth skin. Removal of body hair creates a prepubescent-looking body for women, likely heightening their appearance of youthfulness and increasing sexual dimorphism. In many industrialized cultures, a vast majority of women shave their legs and under their arms for sexy, smooth skin.

Many women also remove pubic hair, despite increased risk for sexually transmitted diseases (STDs) and bacterial infections due to removal of this natural defense. Removal of pubic hair is also thought to decrease lingering pheromones that enhance arousal. When shaving, tweezing, and waxing at home ceases to be enough, many spend money to have hair in these specific areas professionally removed. Waxing is a common procedure in the United States, culminating in a multibillion-dollar business in 2016. Laser hair removal is also one of the most common cosmetic procedures in the United States. Women, in particular, are likely to seek treatments for unwanted facial, underarm, or leg hair, making laser hair removal a $11 billion industry in 2011 in the United States alone.

When surveyed over the last three years, 50–65 percent of women between the ages of 18 and 65 reported that they have regularly removed all pubic hair through shaving or waxing procedures and 22–50 percent reported trimming, grooming, or occasionally shaving. Only between 4 and 16 percent of women chose to let their pubic hair grow naturally. By contrast, only about 20 percent of men totally eliminate their own pubic hair, although most (68 percent) groom or trim with only about 13 percent going natural. Reasons for shaving and waxing vary among individuals, but most report that they are following the cultural norm, that their partner likes it, and women report feeling sexier, cleaner, and more comfortable with their genitals. Women who groom their genitals report feeling more attractive to their partner and report being more satisfied with their sex lives. However, they also report faking more orgasms than women who do not shave. About 24 percent of women prefer a hair-free partner and 60 percent of men prefer a partner who engages in body hair maintenance and removal. The number of individuals who shave their genitals has increased over the last decade, likely due to shifts in availability and trends in pornographic materials. Viewing such materials has shifted what is considered to be normal and has particularly influenced male preferences. Unfortunately, for both genders, shaving or waxing the genitals can lead to genital itching, cuts, and rashes and lowers an individual's defenses against STDs and bacterial infections.

Although women traditionally spend more time and money sculpting their head and body hair than do men, manipulation of head and body hair can also enhance a man's level of attractiveness. For men, darker hair is associated with more maturity and competence. And, in a study conducted in the United States and New Zealand, ratings of male attractiveness were directly correlated with the amount and placement of body hair. At peak fertility, women are more likely to find body hair attractive, but in all other points of their reproductive cycles, women rate men with less hair on their chests and abdomens as more attractive. This research suggests that there are benefits for men who remove hair from their bodies.

The presence of facial hair, alternatively, can be used to increase the ratings of attractiveness for men. Carefully groomed facial hair can increase the appearance of facial symmetry. Research suggests that facial hair can be used to create a more masculine, prominent jaw and the appearance of a more symmetrical face. Facial hair can suggest maturity, high testosterone, and strength. Facial hair can also be used to balance out any asymmetries to create a more symmetrical looking appearance. A carefully orchestrated and well-groomed beard can balance the two halves of an asymmetrical man's face and can make a man appear more confident, dominant, and masculine. Perhaps counterintuitively, baldness can also increase a man's level of masculinity. Bald men tend to be rated as more masculine, more honest, and more dominant by women than men with a full head of hair or men who are balding, as will be discussed further in Part II. Research shows that women in the most fertile phase of their cycle prefer these more masculine-looking men.

There are many different areas of hair on the human body (head, face, chest, armpits, pubic regions, legs, etc.) and each has an impact on human attractiveness, either through its presence or through its absence. Hair creates distinct sexually dimorphic traits, and its placement, distribution, and removal can enhance or detract from one's physically appeal. The typical disbursement of male and female body hair is unique to humans and separates us from other primates. We have less overall body hair than other primates, more hair in specific places (e.g., the top of the head, under the arms, and in pubic areas), and alter and groom our hair. Furthermore, particular hairy features, such as eyebrows and eyelashes, signal the level of interest and pleasure upon meeting others, and raising or furrowing the eyebrows can effectively communicate openness, scorn, or contempt. Eyelashes accentuate and draw attention to the eyes while serving a protective function. Highlighting the eyebrows increases the appearance of youth and overall attractiveness. We can also remove or sculpt hair in particular places to conform to social norms and to accentuate femininity or masculinity.

Thus, to increase individual levels of attractiveness, males and females must pursue different strategies. Women may increase their ratings of attractiveness by emphasizing youthful characteristics. For head hair, this may mean growing longer locks and using lightening products, and using products to increase hair strength and shine. Women may also choose to sculpt eyebrows and remove hair from other places on their bodies. Men may increase their ratings of attractiveness by darkening or shaving their head hair and sculpting their facial and body hair to create the appearance of symmetry. Because hair is such an important feature of attractiveness, the aspects of different types of body hair and their attractiveness and value will be more fully explored in Part II of this text.

Improving the Smile

Given its prominence for communication, the mouth is one feature that is the most noticeable upon an initial meeting. Thus, it is not surprising that the mouth, smile, and teeth significantly contribute to the ratings of attractiveness. The optimal quality of the teeth and smile will be discussed in Part II, but oral hygiene is a highly contributing factor on the ratings of physical attraction. Straight teeth are a signal of good genetic potential. Straight teeth can obviously be a false cue in the age of modern dentistry, but throughout evolutionary time, teeth were a reliable indicator of health. Broken, rotting, or stained teeth, as well as bad breath, are signals of gum disease, infection, and poor health and genetic quality. These traits signal a weaker immune system, which makes the individual more vulnerable to the elements and to disease. As a result of the important communication function of teeth, people from the United States spend billions of dollars every year on dental care, dental products, teeth straightening, and teeth-whitening agents. We also use mouthwash to camouflage or eliminate dental decay or bad breath. Our smiles present ourselves to the world and much effort is put into perfecting them.

Teeth are the backbone of the smile, metaphorically speaking, and research has shown that a smile enhances attractiveness for both males and females. A smile contributes to an open, welcoming expression that encourages interest and engagement. A nice smile suggests stability, humor, and other positive psychological and personality traits. In fact, smile width came in second only to eye size as impacting ratings of female facial attractiveness in a study at the University of Louisville, Kentucky, under the direction of Michael Cunningham.

Those with straight, aligned teeth are rated as more physically attractive, more intelligent, less aggressive, and more socially desirable. As a result of this preoccupation with the smile, 80 percent of individuals in the United States who sought dental treatment in 2015 did so for cosmetic reasons. However, in a study examining the psychological and social effects of orthodontic treatment, Dr. Judith E. N. Albino and colleagues found that the majority of these procedures were unsuccessful in improving the ratings of attractiveness. For example, adolescents who receive orthodontic treatment rate their teeth as more attractive but may perceive their other features more harshly after treatment. Once their teeth are aligned, they start to notice other flaws. So, although orthodontic treatment did align the smile, it was not found to have a significant impact on overall self-esteem and did not alter future behavior. Furthermore, tooth color did not significantly alter the ratings of attractiveness. As long as teeth were not significantly yellowed, teeth-whitening procedures did not impact the attractiveness ratings.

Thus, those interested in heightening initial assessments of attractiveness should make sure to properly care for their smile. Preventative measures such as brushing, flossing, and mouthwash seem to be enough to maximize natural attractiveness. For those with dental problems, more invasive procedures can bolster the appearance of the smile to maximize the initial assessments of attractiveness that are communicated through the smile.

Masking Body Scent

Anyone who has ever been confined in a small space with others knows that scent matters and that it can increase attractiveness or, more commonly, significantly decrease attractiveness. A common way for contemporary people to combat this problem is the use of deodorant and antiperspirants. Prior to modern deodorant, people used scented baths, scented oils (a precursor to perfumes and colognes), and eventually soap. Since the first deodorant was trademarked in the late 1800s, people have been working to make it less messy, less staining, more effective, and more user-friendly. Due to Americans' preoccupation and self-consciousness with regard to odor, use of deodorant has escalated consistently since its invention. In 2017, estimates of the amount spent on deodorants and antiperspirants reached about $18 billion, with sales continuing to increase.

Deodorant is classified as a cosmetic and its initial role was to suppress the growth of odor-causing bacteria in the underarm area. Places such as the warm, dark underarms provide a natural nourishing habitat for bacteria, and their growth produces an odor. Deodorants kill the bacteria that would accumulate under the arms and therefore neutralize the smell. Bathing and soap can also remove bacteria, at least temporarily. Subsequently, antiperspirants were invented to suppress sweating. Although deodorants are typically used under the arms, antiperspirants can be used anywhere on the body to reduce sweat by plugging sweat glands. Since so many people now use deodorants and antiperspirants, the market must keep developing new, more expensive alternatives to keep growing the business. Currently, deodorants and antiperspirants are marketed to include fragrances, novel scents, skin-soothing properties, and more and more effective odor control.

Deodorants can radically increase the ratings of attractiveness, especially when people are in enclosed places. However, it is also suggested that individuals overuse deodorants due to a fear of body odor or due to social pressures. Regular bathing and soap are typically enough to remove bacteria without the use of the chemicals in deodorants, at least for people in cool or climate-controlled environments. Many people are even naturally resistant to the odor-causing bacteria that grow in the underarms. Furthermore, deodorants may be counterproductive and mask one's natural body scent that may be attractive to others.

Individuals, thus, must find a balance that works for them. Deodorant can be highly beneficial for everyone in situations that include groups of people in hot, close quarters. However, finding a partner who is attracted to one's natural scent can facilitate long-term relationship stability and satisfaction and interpersonal attraction. When in doubt, though, deodorant is an easy way to provide a safe first impression when encountering other people.

The Role of Perfume and Cologne

As a nation preoccupied with body odor and deodorant, we know better than most that scent can entice or repel. Although natural body scent may have the greatest potential to allure or deter others, the cosmetics market is rife with perfumes and colognes, marketed with the promise of attracting others. Some actually allude to the situation where users will have to fight the opposite sex off because they will be so aroused and interested.

Research shows that perfume and cologne can actually increase the ratings of attractiveness, particularly in situations where body odor may be a concern. An artificially pleasant scent might mask more unpleasant aromas that would deter others. However, while cologne or perfume can enhance attractiveness, it is important to not overdo it. In our environment of frequent bathing, it may be counterproductive. An overpowering scent can be averse to others or it may mask the scent of one's natural pheromones. Since scent plays a particularly important role for women when they are determining the attractiveness of others, an artificial scent can be misleading and suggest an unnatural chemistry. When the man takes the scent off, the woman may be less attracted to him (or he less attracted to her in the reverse situation).

Thus, natural body scent is key in determining good chemistry between two individuals, so finding someone who finds one's natural scent attractive may lead to the best outcomes.

However, perfumes and colognes do have their place. They can mask body odor in environments where it is unpreventable or when showering is not an option. Different scents can also alter moods, increase sexual appetite, spark memories, calm or excite the senses, and alleviate stress. Choice and quality of scent are important. Specific aromas have been documented to relieve muscle tension, but others have been shown to increase irritability, fatigue, headaches, dizziness, blood pressure, or irritate the skin. Scents can alleviate or cause anxiety and depression. The sense of smell provides a direct route to the brain, which can be beneficial or detrimental, depending on the scent in question.

A review of research findings reveals that good hygiene practices may be more effective than perfumes or colognes to bolster attractiveness. Good hygiene eliminates odor-causing bacteria while still allowing one's natural scent to emanate. When needed, a light, situation-appropriate scent can smooth social interactions and increase initial attractiveness. The key is to find the scent that works with one's natural aroma and to use it sparingly.

Hand and Nail Care

In the United States, women, in particular, put effort into maintaining their hands and nails. Companies that develop and sell lotions, oils, creams, and polishes are multibillion-dollar industries. Many women regularly cut cuticles, file, sculpt, polish, pamper, and visit salons and spas to maintain nail shape and color. Finger length, delicacy, and skin smoothness are the highest predictors of ratings of hand attractiveness, particularly for women. Although finger length is not easily alterable, one can create a more feminine hand via nail care and can smooth skin with lotions. Rings, bracelets, and other adornments can also provide cues of femininity and draw attention to this feature.

Although hand attractiveness did not emerge as similarly important for males, the condition of a man's hands does give information about his level of self-care, his profession, and his attention to detail. Well-maintained, healthy-looking hands communicate good genes, good self-care, and high status. More recently, with the metrosexual movement, men have been given the social license to care for their hands as well. Particularly with rougher jobs and more brute strength, men's hands can take a beating throughout their lifetimes. Manicures and self-care can help maintain hand health and attractiveness even with a rougher lifestyle.

Given this research, it is shown that hand and nail care can increase the ratings of attractiveness, which can influence future opportunities and experiences. The condition of one's hands communicates information to others about his or her aspects of health, social status, and personality. Thus, our preoccupation with hand care aligns with attractiveness ratings, and caring for one's hands, particularly for women, might boost social benefits that accompany higher levels of perceived attractiveness.

Hitting the Gym

In our modern, sedentary environment, the concept of using a gym has become common for staying in shape and maintaining a healthy, attractive body. Attractiveness is highly correlated with health, so working out, increasing cardiovascular health, and building and maintaining muscle tone are key pieces of being attractive.

In 2016 in the United States, almost 60 million people belonged to a gym or health club, making it a $22 billion industry. This demonstrates the value and intention people have to exercise and maintain a healthy body. Unfortunately, 67 percent of these individuals never actually used their gym, so there is a disconnection between intention and practice. Theoretically, joining a

Along with 60 million other Americans, a woman increases her physical health through gym membership. Such activity contributes to increased cardiovascular health, which leads to a healthy oxygenated glow and strong attractive body. Physical activity also increases psychological and emotional health. (Nyul/Dreamstime)

gym can be an effective means to increase one's overall level of attractiveness (and health), but one must actually put in the effort to reap the rewards.

For those who do use the gym or who exercise regularly, there are resulting effects on their ratings of attractiveness. Psychological research shows that women tend to rate muscular, toned men as more physically attractive. Thus, a healthy diet and regular sessions at the gym can increase a male's attractiveness. However, research does show that women prefer toned men to muscle-bound men, so there is no need to overdo it.

Women can also increase their ratings of attractiveness through regular exercise. Exercise increases health, which directly impacts the attractiveness of the body. Similar to males, toned, healthy muscles and increased cardiovascular health contribute to the ratings of females' attractiveness. Well-oxygenated blood gives the skin a healthy glow, and being in good shape lends itself to graceful, attractive movements and body shapes. Such effects contribute to the ratings of attractiveness more significantly than large muscles. The bottom line is that regular exercise contributes to a healthy body, which highly impacts the level of attractiveness.

Resources and the Role of the Wingman

Hitting the gym does serve to enhance a male's physical attractiveness but physical features actually play less of a role in women's judgments of overall attractiveness than they do for men. Therefore, men can enhance their level of attractiveness by enhancing things other than just physical traits. Women are biologically driven to look for a man who is ambitious and who may be a good provider for them and their children. Thus, men can increase their levels of attractiveness by engaging in strategies such as displaying resources or other personal qualities. These may include showing signs of success such as buying sexy, stylish, expensive clothes, watches, cars, or gifts, or by discussing success in business or motivation in a career. Financially successful men are more likely to have more options in dating and relationships, so it benefits them to show off success and display motivation. A well-groomed, well-dressed, tailored male figure is rated as more attractive than a man who is not displaying such traits, regardless of his actual physical features.

Women also use the opinions of others to inform their perceptions and decisions. Men who seem to be fun, engaging, and sociable are rated as more attractive. Men surrounded by friends or who are being given attention by other women are immediately more interesting. Thus, for males, having a friend nearby, particularly a female friend, to provide the illusion of popularity or demand can increase their appeal. Also, as noted earlier, interacting with children or puppies can also provide an immediate character reference.

What women use to make their assessments of male attractiveness also changes with age. Younger women care more about physical attractiveness in their potential male partners. As women age, however, they are more likely to assess personal qualities and use those to assess attractiveness. Attractiveness ratings made by older women are much more likely to be influenced by information about wealth, stability, power, and faithfulness.

Early research about mate selection predicted that signs of success would become less important as women achieve their own occupational success. Once a woman has the power and money to provide for herself without needing a male for support, she should no longer need to rely on his resources to survive. However, research actually shows that a woman becomes even more selective as her own wealth, power, and status increase. Since she no longer requires a male to survive comfortably, she can take her time and choose even more carefully. While a successful male finds that his dating pool deepens and he suddenly has more options to choose from, the opposite is true for women. Women typically choose a mate who is at least on par with her level of socioeconomic status, so as her socioeconomic status increases, her pool of potential partners shrinks, though their quality arguably increases.

Dressing for Success

The fashion industry is currently a half-billion-dollar-a-year enterprise. Clothing, in particular, can be used to enhance one's shape and create an illusion of symmetry and proportion. Just as cosmetics can enhance the face, clothing can enhance the body. Accessories, furthermore, can draw the eye to more attractive features while minimizing attention to less attractive areas. In not so distant history, corsets and hip enhancers were used to catch a man's attention. Today, clothing styles, seam placement, shoes, and colors all embellish the human body to create a presentation that aligns with cultural norms of attractiveness.

As already established, men with broad shoulders and narrow waists, a v-shaped torso, tend to be considered more attractive by others. Although genetics and earned muscle mass are needed to actually achieve this contoured shape, clothing can create the appearance of a more v-shaped trunk. Well-tailored blazers with strong shoulders and a structured torso can create this v-shaped appearance. V-necked shirts can also create the illusion of a v-shaped upper body. An extreme example of creating this shape is seen in the uniforms of football players in the United States. The football shoulder pads create an enhanced and attractive v-shaped torso, but may be difficult to work into one's wardrobe off the field.

Other uniforms can also play a role in enhancing attractiveness, particularly for males. Many professional uniforms are typically cut in a way that enhances a v-shaped torso and also identify individuals as affiliated with particular organizations. By communicating affiliation and status, these uniforms denote power, responsibility, and maturity, and are thus attractive, particularly to women. Uniforms provide information about individuals, such as their roles in society, whether they are military, medical, or part of a gang. Uniforms can identify individuals as belonging to a group or religion; having certain status, ability, class, or function; and as possessing loyalty and achievement. The information a uniform communicates can increase or decrease attractiveness, depending on the communicated role.

Fashion for females is even more pervasive. Clothing can provide the illusion of an hourglass figure, regardless of overall body size and shape. Clothing made for women tends to display more skin, and is typically tailored to highlight the bust, waist, and hip areas. Well-fitting clothing can highlight femininity and can overall enhance a woman's attractiveness. An extreme example would be women's lingerie. Lingerie tends to enhance the bust, cinch in the waist and highlight the hips. This creates the hourglass shape even if it is not naturally occurring, and even when the woman is wearing limited clothing. Furthermore, the feminine cut of suits, dresses, and blouses for women can serve to enhance an hourglass body shape to maximize physical appeal.

Other aspects of women's clothing have also been studied to examine their impacts on the ratings of attractiveness. Examples include the impact of high-heel shoes and the selection of clothing color on the perceptions of others. In the United States, high-heel shoes are typically associated with femininity. High heels cause a shift in a woman's posture and height, accentuate her buttocks and breasts, and naturally highlight the hourglass shape. High-heel shoes also affect a woman's gait and the swing of her hips when walking. Paul Morris and colleagues from the University of Portsmouth examined the impact of high heels on the ratings of attractiveness to illustrate their effect. They found that wearing high-heel shoes caused women to shorten their stride lengths and increased rotation and tilt of the hips when walking. Furthermore, men and women rating the attractiveness of a woman, simply based on her gait, were more likely to rate the women wearing high heels as significantly more attractive than when the women were viewed wearing flat shoes. When viewing just the gait pattern of the women wearing flat shoes, the women were more likely to be misidentified as males. High heels accentuate the sex-specific aspects of the female gait, increasing sexual dimorphism. High heels also increase the contours of the leg and decrease the perceived foot size. Although high heels are usually associated with women, they can be used to increase a man's attractiveness as well. While women's shoe choices tend to be daring and brazen, a covert half inch in the bottom of a man's loafers can give him the height advantage that women tend to find alluring.

With respect to clothing color, Adam Pazda and colleagues from the University of Rochester and colleagues from the University of Innsbruck in Austria examined the impact of clothing color on attractiveness. They found that wearing the color red serves to heighten a woman's attractiveness. Red symbolizes sexual interest and is more likely to catch the eye of a potential sexual partner. The color red is associated with lust and passion and is a typical color for lingerie. In Dr. Pazda's study, the women wearing red were perceived as more attractive and more sexually desirable (as compared to a photograph of the same woman with her shirt altered to be green or white). When examining online dating websites, they also found that the women who expressed an interest in a casual sexual relationship were more likely to be wearing red in their profile photograph. Thus men and women use color when assessing attractiveness and use it to display sexual interest. Specifically, red heightens one's perception of attractiveness and interest in approach by potential partners. Ultimately, Dr. Pazda's research found that men perceived that red clothing signaled sexual attractiveness, sexual desirability, and sexual interest. Concurrently, females tend to wear red when interested in sexual activity and to communicate availability. Men and women may not be consciously aware of these patterns, but their behaviors are likely to be reinforced, influencing future choices and behaviors.

Adorning the Skin

In a quest to alter one's natural level of attractiveness, individuals may adorn their skin with tattoos in addition to their fashion and accessory choices. Although not a recent phenomenon, tattoos are becoming more and more common in Western societies. There have been many studies on the impact of tattoos on assessments of attractiveness and personality for both men and women. In men, tattoos have been found to be associated with risk taking and sensation seeking, group membership (e.g., high status or deviant), and dominance. In women, tattoos are more likely associated with enhancing physical appearance and accentuating femininity. For both genders, Silke Wohlrab and colleagues from the University of Gottingen in Germany expected tattoos to be associated with health, attractiveness, physical aggression, and heightened masculinity or femininity. They found that women did rate tattooed men as healthier and more attractive. However, men rated tattooed women as less healthy and more likely to be promiscuous or to drink alcohol, leading to lowered desirability ratings for a long-term mate. Tattooed male figures were rated as more dominant by both genders, so tattoos for men may have a positive impact on biological and behavioral assessments, but the same was not found for women.

Nicolas Guéguen also found that tattoos did not increase the ratings of attractiveness for women. However, in a field study at a well-known beach, he did find that men approached women with tattoos more often and more quickly. In his study, temporary tattoos were placed on women's lower backs and they were asked to lie on their stomachs reading a book. When the women sported a tattoo, men approached them more often and more quickly than when no tattoo was evident. Although the mean ratings of attractiveness were not influenced by the presence of the tattoo, interest and attention were certainly impacted. Previous studies had found that women with tattoos were rated as being more likely to be sexually promiscuous, which may lend explanation as to why the women were approached in this current study. Regardless of the level of attractiveness, men may be drawn to tattoos as a physical cue of sexual interest and availability.

There are many reasons someone may seek out a tattoo. Men and women may get tattoos for personal reasons, to celebrate an accomplishment, or to commemorate an experience. Tattoos may be used to garner a sense of belonging in a group, as a form of self-expression, or be a by-product of a drunken night out. Research shows that men with tattoos were rated as more masculine and more attractive. However, research shows that women with tattoos were rated as more sexually promiscuous and as more likely to engage in sexual intercourse on the first date. Additionally, research shows that women with tattoos were found to have had more sexual partners including one-night stands than women without tattoos.

HISTORIC METHODS

Cosmetics before Cover Girl

The quest for beauty is not new. The modern cosmetics industry provides relatively safe products for women to enhance their natural beauty but these allergen-free commodities have not always been available. Throughout history, women found other natural cosmetics to enhance their features. Early attempts at enhancing beauty likely seem bizarre in today's society. These attempts at makeup frequently had long-term, debilitating, effects that led to early death. Products that naturally enhanced the eyes, skin, hair, cheeks, and body led to both short-term success and long-term health problems. The trade-off of future health for current attractiveness demonstrates how important the quest for beauty is and has always been. Being attractive to others in youth, when one is reproductively viable, may be worth shortening the life span in an evolutionary sense, contributing to our drive to engage in such practices. Attractive individuals tend to receive more interest from others, have more sexual partners, and produce more offspring. Such genetic success of passing one's genes into the next generation may logically outweigh a shortened life span.

In the Middle East, women used to use powdered lead as a type of early mascara and eyeliner to create the perception of enlarged eyes. The lead darkened the skin and lashes, successfully enhancing the eyes and attracting attention. Unfortunately, the use of lead on the skin also led to metal poisoning and death with extended use. Lead was also used in ancient Rome and 16th-century England to achieve a beautiful, youthful, pale complexion. Users did succeed in achieving clear, pale, youthful skin but then suffered a slow death by lead poisoning.

In Italy, women skipped lead poisoning via mascara and applied nightshade directly to their eyes to dilate their pupils, making their eyes look larger. Unfortunately, the poisonous nightshade significantly damaged their eyes, overall health, and shortened their life spans. Women in England also applied arsenic to their skin to create an unnatural youthful glow. While effective, it also led to early death.

Not all historic skin products were toxic. Egyptians, for example, used crocodile dung mixed with donkey milk as an invigorating facemask to maintain youthful looking skin. While the concoction smelled terrible, it was safer than the other skin creams of the time. In other societies, bugs were crushed and rubbed on the lips to create a rich red lip coloring and mouse fur was cleverly used to create false eyebrows.

Obviously, our ancestors were very creative with many natural forms of cosmetics. In addition to the crocodile dung used in Egypt, the Italians used lion urine to highlight and lighten their hair, while Arabian women discovered a similar trick and used camel urine to make their hair more sleek and shiny. Japanese Geishas used bird droppings as a natural makeup remover to keep their pores clear and create a youthful glow to their skin. English women, in a quest for the ideal hourglass shape, ingested tapeworms to slim their bodies. The tapeworms were quite effective. The worms consumed the food the women ate, successfully creating a trim body. Unfortunately, the women tended to succumb to slow starvation.

Most early cosmetics, including soap, were made at home using whatever was available. Lampblack was used to highlight the eyes and ochre was used to add color to the cheeks. Particularly in male-dominated societies, where women rely on men to provide for their needs, women put effort into appearing worthy of selection for a mate. In these societies, attracting a mate led to better quality of life, more opportunities, and a higher chance a successful reproduction, so shortened life spans may have been an acceptable trade-off.

The Industrial Revolution led to our more modern cosmetic products. Push-up lipsticks and tubes of mascara were invented in the early 20th century.

These inventions started a lucrative trend that quickly led to the production of an abundance of cosmetics, synthetic false eyelashes, modern nail polish, lip gloss, hair dyes, razors, and cosmetic surgery. Money and time were invested into products that could heighten a woman's attractiveness, and the market was ready and waiting.

The History of Tattoos

As in modern times, tattoos were a popular way of adorning the body throughout recorded history. Today we associate tattoos with a wide array of rebel youth, biker gangs, and experimental adolescents, but their roots lie in ancient civilizations. Tattoos were a visible way of identifying cultural belonging and political and social status in small-scale societies. Celts, Greeks, and Romans used tattooing to communicate nobility status, outlaws, and religious observance. Tattoos were used for initiations, to indicate marriage, and to create a visible symbol on those in power for social control. Tattoos were also used to entice sexual partners or to maintain cultural traditions passed on through generations. Today, tattoos tend to be used to identify belonging to a subcultural group, to bond with peers, or as a means of artistic expression, and more and more middle- to upper-class individuals are seeking tattoos. However, originally, tattoos were ways to communicate status, culture, and faith.

The Development of Modern Luxuries

Today, we take the luxuries of hairbrushes, toothbrushes, and the ease of personal hygiene for granted. However, these products were not always available. Hairbrushes were not widely accessible or affordable until the late 1800s. Maintaining long, smooth locks was much more difficult and more unruly hair types made grooming difficult or impossible. This was also the era when toothpaste tubes and modern deodorant were invented. People initially used toothpicks to clean their teeth since toothbrushes were not introduced to the Western world until the late 17th century. During and prior to this era, healthy teeth were more a product of genetic quality than of hygiene practices. Thus, healthy teeth communicated much more than they do today.

Soap was similarly inaccessible or expensive and it was not until the middle of the 19th century that soap became readily available and the middle class began to have the precursors to modern bathroom amenities. Prior to this, baths consisted of sweat baths (rooms full of steam that induced sweating to cleanse pores) or one could occasionally pay to use a bathhouse. Otherwise, baths were predominantly relegated to the summer time when a local river could be used. Middle-class individuals could also occasionally heat water to have a sponge bath or wipe themselves with clean clothes to have a dry bath.

Since bathing is more of a modern luxury, the use of perfumes has been common since prehistoric times. Researchers have found evidence of perfumes being commonly used as far back as 2000 BC in the Middle East. Ancient Egyptians and ancient Greeks were also known to use perfumes regularly in their hygiene rituals. Cologne reigns from the early 18th century when a new scent was created in Cologne, Germany. Although modern use of colognes and perfumes may be counterproductive to finding a compatible mate, in the era of limited bathing, they were likely highly influential in increasing the attractiveness in closed spaces.

CONCLUSION

There are many options used in attempts to increase attractiveness for physical, social, or psychological reasons. Natural methods such as sleep, diet, exercise, facial expressions, and posture do serve to enhance natural beauty. Cosmetics, hair dyes, lotions, fashion, and hygiene products can highlight qualities, create the illusion of symmetry, or hide flaws. Surgery can artificially alter naturally occurring characteristics to create symmetry, ratios, or more attractive features. Physical changes can create a positive feedback cycle and increase confidence, enhance self-esteem, and alter behavior. These behavioral changes can put forth a persona that is more confident, outgoing, and attractive to others. This attraction can lead to more positive interactions, more opportunities, and a better quality of life.

The attempt to improve one's physical appearance is not a new endeavor. Early Greek and Roman art and records indicate ubiquitous attempts to increase attractiveness. Today, we still see remnants of the same behaviors of cosmetic use, fashion accessories, and clothing choices, as well as the use of colognes and perfumes to ensnare the senses. Men and women tend to attempt to enhance their own attractiveness even if they are not consciously aware of the myriad of benefits such changes bring.

4
Changes in Beauty Trends over Time

Secondary to biological drives and evolutionary pressures, research shows that social pressure, predominantly communicated through media, is the biggest driving force behind which traits are considered to be most attractive. In the United States, in particular, mass exposure to television, movies, fashion magazines, and the Internet provides a breeding ground for pervasive and transforming societal expectations. The media creates norms, drives fads, and presents digitally created perfection. Media determines and displays current trends, impacting ideal characteristics. Fluctuations over decades put different pressures on generations of young women and men, shifting cultural norms. Due to this, grooming, dieting, and self-care practices change from generation to generation. While there are some generational consistencies in ideals for women and men, the ideals for many features constantly shift, reflecting the changes in society. For example, waist-to-hip ratio (WHR) ideals for women and the preference for taller, stronger men have not changed much over the decades, but hairstyles, fashion choices, and ideal body size are modified with each generation.

These social expectations can enforce contradictory goals as fads change. For example, at some points, pale skin was the height of beauty; at others, a bronze tan was sought after. Ideal weight, ideal fashion, and ideal physical traits have shifted throughout time within and between cultures. Research shows that what we find attractive tends to line up with what is present in our culture and we seek to emulate what we know. In a culture informed by Hollywood movies, television, and fashion magazines, what we see are tall, slim, busty women and muscular, masculine men, creating that expectation for all individuals. Prior to widespread television, such emphasis on physical perfection was not as common and individuals tended to compare themselves to family, peers, and friends.

Although which traits are considered to be the most attractive have shifted, those ideals tend to vary predictably in relation to societal norms for different levels of socioeconomic status. As characteristics of what it means to be wealthy change, so do characteristics of beauty. For example, in cultures where the poor must work outside, constantly exposed to the sun and who are thus darkly tanned, and the wealthy have ivory complexions from staying inside and pursuing lives of leisure, lighter skin is prized. In cultures where the poorer and middle classes work indoors at desks or in factories and do not have the luxury of spending time lounging outside, a bronzer complexion is sought after. Similar to ideal complexion, ideal weight varies depending on indicators of status. In times of drought or famine, heavier men and women are considered more attractive. In times of plenty, the thin are the most beautiful. In all cases, however, beauty is tied to status, social expectations, and usually reproductive potential. This chapter explores changing beauty trends for women and men over the last century.

CHANGES IN BEAUTY IDEALS FOR WOMEN

Ideal beauty for women over the previous centuries has shifted from an emphasis on personality and behavioral attributes such as tidiness and propriety to a reliance on physical traits. Such shifts have led to higher rates of vanity, self-consciousness, and greater emphasis on sexuality. Personal identity for girls used to revolve around skills, achievements, and personal characteristics. Now, identity for girls, beginning as early as grade school, revolves around body shape, size, and physical presentation. As characteristics of beauty have shifted, identity development has been impacted. Prior to the 20th century, goddesses were revealed as female forms with wide hips and full breasts. A plump female figure was celebrated and natural. Social pressures encouraged women to embrace their full figures and the life that they provided and sustained.

With the emergence of paper advertisements in the late 19th and early 20th century, advertisers found a forum to influence the population to sell more products. Since sex is enticing and attention grabbing, female models began the slow trek of becoming more sexualized and idealized. By the early 1900s, women were informed about how they should look, be shaped, and dress from advertisements and magazines, and society saw a marked increase in eating disorders such as anorexia. In this way, culture has shaped what a woman believes she is supposed to look like. Women responded by dieting and purchasing the bras, corsets, slips, and girdles to achieve these new cultural ideals. Corsets functioned by narrowing at the waist, emphasizing the breast and hips and helping a woman maintain posture. Unfortunately, the girdles restricted body movement and physiological processes and led to health problems, ultimately leading to a movement to "burn the bra" several decades later.

In recent decades, in the Girl Scouts, there was an attempt to shift attention back to accomplishments rather than looks. But even then, Girl Scout troops were asked to have girls' measure and track changes in their height and weight, and outward appearance was still emphasized. Culture was both telling girls that they should cultivate beauty and skills within while enforcing physical expectations for their outward appearances. A new campaign within the Girl Scouts to improve body image for young girls recognizes the impact of media and societal pressures on girls. To combat this impact, the Girl Scouts have partnered with Dove and have launched a campaign called "Changing the Face of Fashion." The focus of this campaign is to expose girls to more diverse role models to increase inclusion, representation, and confidence rather than emphasizing physical appearance.

Beginning in the early 1900s, the feminine ideal was a petite woman with a boyish figure and feminine makeup. Straight flapper dresses were the rage, accentuating a long thin body. Women responded by a new focus on thinness and an increase in dieting. By the 1930s, this ideal was replaced by a preference for curvy women with a strong silhouette. Dresses that accentuated the feminine figure emerged. By the 1940s, wide shoulders and an emphasis on long, toned, legs were in style. Fashion icons began to incorporate shoulder pads that enhanced a woman's shoulders, and nylons were introduced to smooth the appearance of long legs. Miss America contestants reflected the changing standards and winners grew in height; shrank in weight, waist size, and hip size; and grew in bust over the next 40 years. Pinups were more fleshy and voluptuous while fashion models grew more and more slender.

By the 1950s, female bodies emphasizing full hips and large breasts with a small waist became popular. In the 1960s, petite women with boyish figures cycled back into style. In the late 1960s, shockingly slender became the rage. The entrance of cultural icons such as the Beatles brought British fashion and miniskirts to the American culture. These fashions highlighted slender hips, and slim legs were emphasized. Miniskirts emerged and curves were minimized in preference for small, trim bodies. In the 1970s, the ideal again shifted and large breasts and narrow hips became popular. By the 1980s, wide shoulders and heavy makeup were back in style. Also, beginning in the 1970s, men's magazines introduced more common inclusion of female genitalia and the focus shifted from above the waist to below it. By the 1980s, when surveyed, men chose the butt over the breasts as the sexiest feature, more than two-to-one.

To reflect the fashion of the 1920s, women highlighted long, thin torsos with straight flapper dresses, and wore feminine makeup. This fad led to a focus on female thinness and dieting that lasted until a more curvaceous ideal emerged in the 1930s. (Library of Congress)

A cultural movement toward women's participation in athletics gave women an opportunity to see value in viewing themselves as powerful and aggressive, in high opposition to their mothers' era. Moving into the 1990s and beyond, a preference for excessive thinness and large breasts emerged, promoting the increases in surgeries for breast implants and the incidence of psychological disorders such as eating disorders. The 2000s brought extensive cosmetics and an explosive growth in cosmetic surgery and even thinner bodies. Only recently has a shift to toned, muscular bodies begun, with less emphasis on thinness and more emphasis on health. Although this shift could potentially be beneficial for women's psychological health, it also adds another layer to what women are supposed to be. Now ideal women are thin, curvy, and muscular, a demanding ideal.

Changing Fashion Highlights Changing Ideals

Changes in beauty ideals for women are correlated with changes in the fashion industry. Rather than to simply warm and protect, clothing serves the additional purpose of enhancing or hiding features of the body. Traditionally, fashion served to hide a woman's body, even for swimwear. Bulky long skirts may not be helpful once weighted down with water, but they did provide modesty for the hips and legs and kept the skin pale and untouched by the sun. Women even sewed weights into the hem of their bathing gowns to prevent them from floating up and revealing their legs. With the addition of bonnets, shawls, and gloves, American women in the 1800s were ready to spend a modest and anonymous day at the beach. Later, shorter gowns with loose flannel pants gave women a little more functionality when swimming while still preserving modesty. Although the clothing was still heavy and cumbersome, it was a step forward in accessibility. Later, wool dresses and long stockings and slippers became the fashion for a day at the beach.

Entering into the 1900s, swimwear began to actually reveal a female body underneath the layers of fabric. By the 1920s, one piece swimwear emerged and allowed women to actually swim rather than just stand in the waves. This swimwear resembled a long top over shorts and stockings and, for some, skin could be seen between the hem and the stockings. The tendency to be more revealing has continued over the decades and one can look at current beaches to see how far this trend has come. As bathing suits, and clothing in general, have become more revealing, there is more judgment of feminine attractiveness for a range of body parts. A boyish body may have been the rage in the early 1900s because that is what clothing allowed a viewer to see. Today, with a vast proportion of the body exposed in clothing of all types, there is more for others to judge.

Researchers question whether beauty ideals follow the fashion trends or if fashion changes to keep up with new ideals. It is likely a little of both. In the mid-1800s, for example, a small waist was a hallmark of femininity and attractiveness. This trend was so extreme that corsets were used to create a narrow waist, even at the expense of a woman's health. Tight corsets compacted organs and impeded attaining adequate oxygen when breathing, leading to faintness and weakness for the most fashionable women. In this time, known as the Victorian era, frailty actually became a hallmark of femininity. Women who were weak, pale, and in need of protection were heralded as the hallmarks of femininity. Luckily, by the 1880s, a heavier, durable frame became the fashion. Women worried about being too thin and actually used padding to create the perception of greater bulk.

In pre–World War I years, ideals ranged from curvy and voluptuous to slender and delicate and fashion trends reflected these ideals. Further into the 1900s, slender but strong became the ideal. Corsets again were used to create a slender waist and defined bosom and hemlines shorted to show a hint of female legs. Into the 1920s, after World War I, the ideal garment lost the cinched-waist look and the ideal young female had a boy-like frame with narrowed hips. Corsets were abandoned and the hallmark of beauty shifted to the legs and face. In 1930, hemlines lengthened, and the narrow waist returned, curves were accentuated and breasts were hailed. Flat stomachs became the rage and small firm breasts that stood as individuals rather than as one corseted mass were ideal. In the 1940s, nylons accentuated the legs and butt. Following the emphasis on the lower body, the breast again began to swell. The preferred size swelled from a flat boyish body in the 1920s to voluptuous proportions in the 1960s. By then, large cleaved breasts, tiny waists, and wiggly walks were created and accentuated by fashion choices.

Because men place more emphasis on physical attractiveness than women, women's fashion has been more extensive and more variable over generations. These changes show that, although beauty standards are largely universal, ideals do vary over time and across cultures. Fads, ideals, and advertising shift social standards. These changes include the steady trend toward smaller waist size for women and eras of "bosom mania" when women worked to accentuate the bosom either through clothing, exercise, or surgery. Many societal trends result in healthy women who are dissatisfied with their bodies and who go to great lengths to conform to societal pressures. Furthermore, due to increased attention on physical appearance, women, arguably, have to work harder to maintain standards to secure professional and relationship opportunities and to be socially acceptable.

Changes in Beauty with Age

Although changes over the decades have influenced what is considered beautiful, there are similar changes in what is considered to be ideal beauty over an individual's life span. The human body goes through typical changes in the aging process, many of which are at odds with what is considered ideal for attractiveness. Since attractiveness is closely tied with fertility, women in their twenties and thirties are usually considered to have qualities that are rated as the most attractive. Thus, female adolescents may endeavor to make themselves appear older, such as using cosmetics, shaping eyebrows, and making careful fashion choices to align with current trends. Women in their twenties and thirties tend to be the most focused on physical appearance. Most are fertile and interested in finding a partner or procreating, thus spend time maintaining their appearances, as they are concerned with physical appeal.

Older women tend to be rated as less attractive with age. These women grow less fertile over time, and less metabolic energy is wasted on maintaining physical features that advertise reproductive viability. Thus, with age (as in childhood), female bodies more closely resemble male bodies, with less definition between the waist and hips. The change in physical appearance can be detrimental to a woman's self-esteem but it can also allow her to value and be valued with respect to other traits. Many older women are physically and socially active and play a strong role in the community or the family, such as helping with grandchildren. They may be leaders in business or volunteer groups, and assertiveness and wisdom may be more valued assets than mere physical beauty. Many older women are also in long-term relationships where their value as a partner is no longer simply a by-product of their physical appeal.

Since typical beauty is correlated with youth, to maintain their physical appeal, many aging women also expend great effort to maintain a youthful appearance. In an interview study carried about by Enguerran Macia, Priscilla Duboz, and Dominique Cheve, from the University of Bamako and Aix-Marseille University, aging women reported being very aware of the societal and cultural pressures to stay young and thin and considered their aging bodies to be ugly. Eighty-two percent of women recognized the importance of maintaining their appearance while only 60 percent of men felt the same pressures as they aged.

The aging body may become a stigma for women, and aging is revealed through wrinkles, less supple skin, graying and/or thinning hair, sagging muscles, and increased weight around the midsection. Beauty products such as skin creams, hair dyes, body oils, and vitamins are increasingly used with age, putting an increased expense on older women. As in youth, beauty is used as a signal for moral character, and older adults are judged more harshly when they appear their ages or older. Those who have the financial stability to do so are more likely to turn to cosmetic surgery to maintain youth. Many successful older adults embrace the concept of inner beauty, denounce surgery (especially if they do not have the means), and find ways to interact with and have an impact on the world without relying on their looks. Most look back on photos of themselves and realize they judged themselves too harshly when they were young and wish they had realized earlier that they were beautiful.

As mentioned, attractive older adults tend to be those who still have characteristics of youth. They have shiny healthy hair, smooth skin, trim figures, good muscle tone, and clear eyes. Genetics can definitely impact the aging process and the cosmetic industry can help fill in the gaps. Older adults will find that they cannot prevent the aging process, and, with time, external beauty fades despite attempts to delay the process. Unfortunately, as a culture that values youth and beauty, older adults in the United States tend to be pushed aside for newer and more beautiful models. The United States may be well served to emulate other cultures, which value their elders, understand the wisdom that comes with age, and maintain healthy respect for older adults even as beauty fades.

Consequences of Beauty Ideals for Women

Beauty ideals for women have clearly cycled over the previous generations. At a psychological level, these competing ideals can be exhausting and damaging to women's self-esteem, as will be discussed in Chapter 6. Currently the unrealistically thin ideal for women has contributed to a rash of eating disorders that were very rare a few decades ago. Furthermore, these superficial, haphazard ideals are missing the evolutionary point. There is a biological balance for healthy reproduction. A woman who is too thin or too heavy is not reproductively viable. A weight and shape that is healthy for the individual woman's body carries with it the most reproductive success. Chasing cultural ideals means bowing to societal rather than evolutionary pressures. In this area, research suggests that culture and media may be doing women a disservice. The era of Photoshop and perfection is creating an ideal that is difficult to achieve and detrimental to future health and reproductive success.

Today, women diet to a degree that is counterproductive to their health. Women who tend to win beauty pageants and modeling contracts are actually underweight and are risking their future health and reproductive potential. Some countries, such as France, have instituted controls to not allow the perpetuation of unhealthy ideals. For example, models in France must acquire a doctor's note to confirm they are a healthy weight. Furthermore, if photographs are retouched, that information must be noted on the photo. Such measures are good first steps to help women who are dissatisfied with their bodies due to unrealistic ideals set forth in fashion. When interviewed, almost all normal-sized women endorse feeling that they are too fat, their butts are too big, and their breasts are too small, particularly after viewing a media image of a female model. The damage to their self-perception is so vast that cosmetic surgery has become a multibillion-dollar industry to enlarge breasts, suck out fat, and reshape features to chase unrealistic ideals.

Some companies have recently made an attempt to reverse this psychologically damaging trend in the advertisement industry. For example, the Dove campaign uses nonidealized models to advertise its brand in an attempt to solicit consumers who are more likely to identify with more normal-sized women in the ads. Interestingly, Dove has had some success in its marketing campaigns, but its success is contingent upon the level of self-esteem of the consumer. Michael Antioco and colleagues from the Emlyon Business School in France found that consumers with low self-esteem reported being more receptive to and more interested in products being advertised by more averaged-sized women. Consumers with low self-esteem were, similarly, less likely to purchase products advertised by thin models. Those women with high self-esteem report being more likely to purchase products advertised using idealized models. This increase in purchasing may be due to their high self-esteem contributing to the belief that such perfection is attainable. However, the normal-sized models were perceived as having increased levels of trustworthiness by all consumers (those with high and with low self-esteem), which generated increased expressions of intention to buy the products being advertised by all consumers.

CHANGES IN BEAUTY IDEALS FOR MEN

Just as has been the case for women, shifting societal pressures have an influence on what is considered to be attractive for males over time. These differences are retroactively predictable because most directly correlate to characteristics that are indicative of wealth and status. Early on, however, philosophers thought that golden proportions and perfect symmetry were the fundamental keys to ideal beauty. In the 1400s, Da Vinci attempted to design this perfect male figure in his Vitruvian Man. This figure was based on mathematic calculations and architectural theory and specified ratios of height, width, lengths, and proportions for essentially every body feature. Although these proportions have not panned out to create a universal ideal, they are still prominent in modern research. Research paradigms are built around exploration of ideal shoulder-to-hip ratio (SHR) and WHR. Both factor into physical beauty and researchers are still striving to find the ideal that summarizes the perfect proportions of ideal beauty from a cross-cultural perspective.

However, evolutionarily perfect proportions are arguably not as important as health and reproductive success, although proportions and health do tend to be correlated. Proportions that support survival and reproduction within a given environment are the ones that tend to be more readily passed down to greater numbers of offspring. Research has demonstrated that muscular males tend to have the underlying metabolic and genetic health to support

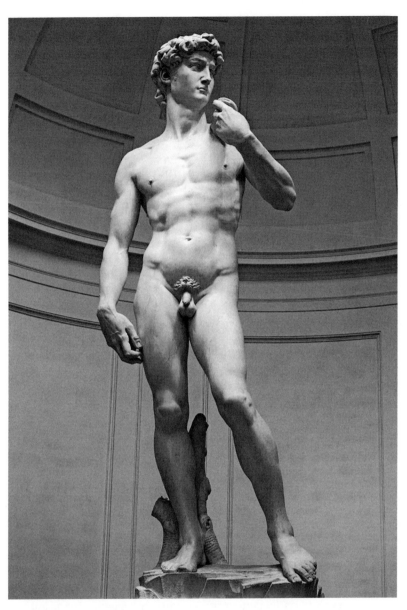

Michelangelo's marble statue of *David* (1501–1504) displays a youthful, symmetrical, lean muscularity, and a masculine shoulder-to-hip ratio that reveals health and an enduring physical appeal. (Wilson Guariglia/Dreamstime)

the metabolically demanding muscle tissue and risk-taking behaviors. These males also advertise the likelihood of greater reproductive success due to maintaining healthy bodies and having increased abilities to protect partners and offspring and intimidate other males. Strong males are able to work, hunt, and provide, as well as protect.

As early as the ancient Greeks, the ideal male form displayed well-defined muscles, low body fat, and aesthetically pleasing proportions. Physical activity and training characterize all of the great heroes of the time (e.g., Achilles, Alexander the Great, and Julius Caesar). During the medieval stages, poor men were more likely to be underweight, have poor muscle tone, and be wasting away. The wealthy had the luxury of food, physical exercise, and good health. Thus, well-defined muscles and healthy proportions had the dual role of indicating health and status. Even into the Renaissance, Michelangelo's *David* displayed the qualities of being youthful, symmetrical, and lean, yet muscular. Alternatively, in ancient Egypt, the ideal male had a radically different body type. The wealthiest men were careful to stay lean, care for their skin and bodies, and keep themselves clean by shaving and washing. Muscularity was saved for the workingmen and soldiers whose job was to protect the royalty who did not need to waste time on such pursuits. Thus, a slender body indicated wealth and status and was sought after as the ideal. Similarly, in Roman art of the same era, young, undeveloped males were held as a standard of beauty. These males were slim, nonmuscular, and fair, contrary to today's masculine ideals.

By the 1800s, the ideal American male differed from either of these earlier ideals. Due to uneven distribution of food and resources in the United States, the most attractive males were portly with wide waists and sturdy legs. The ideal male during this time would be considered overweight by today's standards. Given this radical physical difference, in the late 1800s, the most sought after SHR would have been drastically different than it is today, with ideal male hips rivaling the width of their shoulders. Such proportions communicated wealth and status since only the most successful males had access to excess food and leisure. In the late 1800s, an elite club was actually formed in Connecticut that only admitted men over 200 pounds. By the turn of the century, however, the club closed because social policies and ideals were changing and heavier men no longer held the same status.

By the early 1900s, preferences had shifted and a slender build was considered more ideal. This aligned with societal changes that made food more accessible to more people. Heaviness was no longer indicative of success and status. Furthermore, more Hollywood stars, such as Cary Grant, Jimmy Stewart, Fred Astaire, and Gene Kelly, were modeling trim-toned figures. By the mid-1900s, counter-cultural ideals emerged and rebellious, long haired, slim, lanky men were heralded as more attractive, on average. These men were often sons of the wealthy and did not need to spend time working, exercising, or making money. They appeared to be less driven and spent more time at leisure, thus they tended to have thinner, less muscular, builds.

By the 1980s, bodybuilding came into fashion and male norms and ideals shifted again. Hollywood actors led the way, and actors such as Arnold Schwarzenegger and Jean-Claude Van Damme illustrated remarkable muscular extremes and sported body frames unattainable to most healthy males. Male power soared and men spent more time working out to bulk up their shoulders and arms. After hitting its peak in the late 1980s, the male ideal shifted again to more slim and toned. This was good news for the average male who may not have the time or energy to chase the muscular bulk of the 1980s. More recently, however, an increased preference for muscularity has reemerged, so men seeking the ideal body shape may need to schedule more time at the gym once again.

Current perspectives on male attractiveness include an increase in muscle size, bulk, and definition. For males, rather than thinness, the ideal is to be highly muscular. The males who are rated as the most attractive by women, as well as by other men, tend to be more muscular than the average male. Well-developed muscles of the upper body, such as the chest, arms, and shoulders, contribute to a high SHR in a male body and seem to be particularly desired. In current psychological research, the SHR accounts for the largest proportion of variation in attractiveness. SHR is more influential than body mass index or WHR for levels of male physical attractiveness.

Media and social influences remain a driving force for physical ideals. Models and actors have become more muscular and masculine over time. Even toys have become more muscular with the most recent G.I. Joe doll rivaling the largest human bodybuilders. Unsurprisingly, this is an unrealistic goal for most males and we are exposing our boys to these pressures at very young ages. As a result of these pressures, research shows that men choose a preferred body type that has about 8 percent less fat and 25 pounds of additional muscle than what they actually have. Lean muscularity is the new Western male ideal. Now that women hold similar positions of power, muscularity may be one way for men to maintain superiority over or distinction from women and to catch a woman's attention. Also, now that women have their own sources of income, they can be more selective about their partners and mates and have the luxury of choosing physical features, rather than just seeking provisions and protection.

For both genders, body shape highly influences physical attractiveness. The ideal body shape, however, is a moving target and what is ideal today will likely change over the next decades. In addition to body shape, many other male characteristics also contribute to the ratings of attractiveness as well. These include features such as cranial hair, beards, body hair, skin color, facial features, height, and fashion sense, just to name a few. The quest for attractiveness, it seems, is a rigorous enterprise for males as well as females.

Changing Hairstyles and Fashion Highlight Changing Ideals

Other characteristics aside from body shape and muscularity, which have shifted over time, include male hairstyles and fashion trends. In the early 1900s, slicked back hair and a handlebar moustache were the rage. In the 1930s, fashion trends shifted to a thin moustache and sculpted hair. In the 1940s, clean-shaven came into fashion. In the 1950s, slicked, Elvis-style hair was rated as most attractive for the modern male. In the 1960s, men grew their hair long and there was a movement toward anti-disciplinary grooming and shaving. A more hippy-like appearance came into fashion with a de-emphasis on maintaining a strict appearance.

By the 1980s, this fashion waned and men returned to slicked back hair and a more professional look. But, in the 1990s, there was a regression again toward a grunge-rock look. More recently, there has been an emphasis on a more ultra-groomed looked and metrosexuals became common in the early 2000s. Metrosexuals spend more time and effort on their appearance and may be seen sporting frosted tipped hair and using skin products to soften and pamper their skin. Finally, the lumbersexual has emerged within the last decade. These are men who cultivate the appearance of rugged facial hair, muscularity, and flannel shirts despite urban lifestyles. For these shifting styles, cultural ideals helped shape what was considered attractive independently of ideals shaped by evolutionary pressures.

Beards, in particular, have had eras when they were associated with masculinity, dominance, courageousness, and confidence. Beards are a male symbol that can be associated with power, strength, and virility. Particularly during the 1800s, beards were associated with social power and political leadership. Abraham Lincoln used a beard to strengthen his facial features and increase his image of strength and wisdom. Entering into the 1900s, media started showing clean-shaven men and there was a shift to beards becoming less popular. A shaved face may be rated as more hygienic, friendly, welcoming, and engaging and beards started to become associated with deviant groups. More recently, the trend of enhancing masculinity and attractiveness with carefully groomed facial hair has returned, with the emergence of the modern lumbersexual.

Fashion trends have been similarly revealing of cultural ideals throughout the generations. Although today ruggedness is considered to be masculine and attractive, in Europe in the late 18th century, men actually wore corsets to give themselves an effeminate look. Similar to the early Romans, a more delicate body type correlated directly with societal status and was most common in those individuals who had wealth and power and did not need muscles to defend themselves against enemies. This trend continued into the 19th century and such men became known as "dandys." Although this term may hold a negative connotation for today's males, throughout that century, lean, slim men were the most prominent, powerful, and attractive, and muscles were saved for the working class.

In the 17th century, men of status also used to wear high heels. These were primarily used by men to help steady themselves in the stirrup on horseback when standing to shoot arrows or fight. They had the added effect of making men appear taller and more attractive. Heels became another indicator of status (even today, the term "well-heeled" indicates someone of wealth or status). Since heels are not conducive to walking on uneven ground or working in fields or laboring, this fashion trend created another visible distinction between the classes. Over time, women started wearing heels to show their equality and eventually men moved away from heels to show their distinction from women.

Although today the ideal male is slim and muscular, the variations across time are understandable when viewed in light of the socioeconomic climate of the times. Plumper, less muscular men were viewed more positively in times when food was scarce and the average man needed to engage in physical labor to survive. Such labor led to increased muscles, so such physical traits displayed signs of being poor or part of the working class. Plumpness in this era indicated privilege, wealth, and the luxury of food and relaxation. Since signs of social status are appealing to women, these physical indicators of wealth and status would increase attractiveness, independent of genetic traits.

Changes in attractiveness ideals over history clearly demonstrate that cultural pressures can shift what traits we find attractive and how we choose to present ourselves, even in opposition to what may be predicted by evolution. For example, the metrosexual of the late 1990s combines caring for one's body, caring about fashion, and maintaining good grooming and hygiene habits. This is the opposite of the evolutionarily predicted hairy, muscular, male ideal. Although evolutionary pressures would suggest that a hairy male would demonstrate having more testosterone, would be seen as more aggressive, and would be more attractive to women, by 2012, 60 percent of males from the United States and Australia reported engaging in body hair removal. This change in behavior directly relates to cultural pressures rather than evolutionary ones.

Consequences of Beauty Ideals for Men

Although the effects of extreme beauty standards for women are more frequently highlighted in research, attractiveness standards for men carry consequences as well. In modern times, movies and television sculpt the cultural standard, from the lanky frame of Fred Astaire in the early 1900s to the rugged masculinity of John Wayne in the 1960s and to the superhuman Marvel superheroes of 2010s. Cultural, economic, and social factors have an influence on what is ideal, fashionable, and attractive for men just as they do for women. Although a wide variety of body shapes and sizes can be healthy, the media currently highlights excessive musculature as an emerging measure of attractiveness, particularly over the last few decades. For men whose careers are not built around sustaining muscle size and tone and whose genetics do not support such body shape, such proportions are unrealistic to attain and maintain, and impact male self-esteem just as readily as ideals for women.

The focus on lean muscularity for today's American male has led to an increase in eating disorders, decreases in self-esteem and body image, and an increase in the number of men seeking cosmetic surgeries. Although cosmetic surgery is still more common for women, more and more males are seeking cosmetic procedures, particularly liposuction to decrease fat around the middle, to heighten their SHR. These procedures are simply for cosmetic reasons and carry the risks associated with surgery. Unrealistic ideals also influence male psychological health and stability, and although there has been much research on the negative effects of media for women's self-esteem and body concepts, there is now a need for similar work for male perceptions.

CONCLUSION

Although the basis of attractiveness is biologically predicted and fundamentally innate, research shows that there is room for environmental pressures to alter which characteristics are perceived as physically attractive. Furthermore, these traits can change over time. Throughout generations, ideal characteristics have shifted with environmental, socioeconomic, and media pressures for women and men. Basically, those traits associated with high status tend to emerge as the most attractive. This can be seen in examples of preferences for skin tone, body weight, and muscularity for women and men throughout the decades. Furthermore, the media has the ability to further shape ideal body characteristics. Those traits that are featured in magazines, television, music videos, and movies become the ideal for each gender. Those in the limelight create the cultural norms for body weight, muscularity, and physical features, regardless of true averages. Viewers take those characteristics as a guideline for what is normal and attempt to change their own bodies accordingly, whether or not such changes are realistic.

Conventionally, the media uses sexual appeal to sell products and to grab the attention of viewers. Advertisements for diverse products such as cars, jeans, and gum all take advantage of a beautiful image to engage and entice consumers. These images influence gender expectations and norms at the same time. While the advertisements may serve to capture attention and sell products, the impact of such imagery has had negative effects on overall body image and has increased the extent of body-consciousness of modern society. Rather than only valuing health and the reproductive fitness, modern women and men engage in extreme dieting and cumbersome bodybuilding and are self-conscious about the size of their bodies and muscles. Unfortunately, since sex appeal does attract attention, consumers are compounding the issue. By purchasing the products, consumers are reinforcing the unattainable ideals being used in advertising and contributing to their own body-consciousness.

Thus, the media has the power to shape the physical ideals of a culture. Since being attractive is correlated with physical, social, and psychological benefits, being unable to meet unrealistic ideals leads to decreases in physical and mental health and stability in the population, as will be discussed further in Chapter 7. Although underlying genetic preferences for symmetrical and youthful characteristics are still present, culture can shape the exact characteristics and proportions of the ideal female or male body. These ideal characteristics are revealed by variations in hairstyles, skin tones, cosmetic usage, surgical procedures, dieting, exercise habits, and fashion choices over generations.

5

Evolution's Impact on Modern Attraction: The Interaction of Genes and Environment

Chapter 4 illustrates how changing cultural norms have contributed to changes in attractiveness for men and women. Media, fashion, and socioeconomic pressures have influenced which features are rated as the most attractive over time. However, although preferences for some features have changed over the last century, other preferences seem resistant to such changes and have remained consistent and predictable. This consistency can be explained by considering the evolutionary pressures that shaped modern humans. This chapter examines how our evolutionary history has shaped our ratings of attractiveness and how differences in our modern environment predictably impact what individuals find to be attractive. Research shows that, across cultures, preferences vary predictably depending on gender, type of relationship under consideration, and socioeconomic climate. Thus attractiveness is a by-product of the interaction of our genes and the environment in which they are expressed.

First, this chapter addresses why there tend to be gender differences in what men and women find to be the most attractive and the consistency among cultures. Second, differences in the ratings of attractive traits are examined with respect to different types and lengths of relationships, such as friendships, short-term romantic relationships, and long-term relationships. Third, the impact of environmental circumstances such as economic climates, sex ratio, and exposure to pathogens are addressed. Although the foundation of attraction may be genetic, as discussed in Chapter 1, these genetic preferences interact with our modern environment. This interaction explains why different characteristics are rated as more or less attractive given different circumstances.

AN INTRODUCTION TO NATURAL SELECTION AND SEXUAL SELECTION

The traits one finds attractive are fundamentally a by-product of the forces of natural and sexual selection. Natural selection suggests that those traits that help an individual survive to a reproductive age will be more likely to be passed on to offspring. Individuals who possess traits that do not help them survive are less likely to live to pass their traits on to the next generation. Over evolutionary time, the traits that allow individuals to survive expand in the population and those that do not contribute to survival are weeded out, changing the genetic makeup of the population as a whole (i.e., evolution).

Sexual selection is a special form of natural selection that is focused on those traits that make an individual more likely to be chosen as a reproductive partner. Being able to survive to reproductive age is a good start, but it is not enough for traits to be maintained in a species. Individuals must successfully reproduce to pass on their genes to the next generation. So, individuals must have the traits that allow them to survive to reproductive age and traits to attract a sexual partner to pass on their genes.

The traits needed for survival and the traits needed to make one competitive for finding a reproductive partner may align or they may have competing effects. The results of such competition can be seen throughout the animal world. Males and females of many species look radically different due to the pressures of sexual selection. In birds, for example, camouflaged feathers are highly conducive to hiding from predators and for surviving to reproductive age. In direct opposition, however, brightly colored feathers are conducive to catching a partner's eye for successful reproduction. Because of sexual selection (and differences in parental investment, which will be discussed next), we see extreme physical differences between males and females. Females have evolved to be camouflaged and blend into the environment, while males have evolved to be brightly colored. The colors are a double-edged sword because more colorful male birds are more likely to successfully reproduce but they are also more likely to be eaten by predators.

Similar dichotomous traits are apparent in many other species. Male lions are so big and bulky that they are not able to hunt effectively like their slender female counterparts and thus males are more likely to die if not part of a pride. The bulkiest males, however, can intimidate and outcompete other males and reproduce with many females, making this trait beneficial for reproduction, if not for long-term survival. In humans, males tend to be more physically aggressive and have higher rates of cell-damaging testosterone.

The more aggressive males are not likely to live as long as less aggressive males and are significantly more likely to die due to risk-taking behaviors, but they are also much more likely to successfully reproduce than their less aggressive counterparts, providing an evolutionary trade-off to sustain the trait in the population.

Similar to other species, sexual selection explains many biological and behavioral differences in men and women. For successful propagation of the human species, not only must men and women survive to reproductive age, but they must also display traits that make them competitive for selection as a reproductive partner. Those individuals who display the most attractive traits are the ones who are more likely to reproduce, and hence pass those traits to the next generation. Successful females include those who display traits that signal reproductive viability (wide hips, youth), and successful males display traits that signal ability to provide (physically, financially, and emotionally). These differences in ideal traits directly contribute to the reproductive process. Men and women play radically different roles in reproduction, particularly with respect to parental investment, and thus are subject to different environmental pressures.

THE ROLE OF PARENTAL INVESTMENT

Parental investment refers to the amount of time and energy a parent provides when raising children. In humans, as well as most other mammals, females tend to invest more time and energy in raising children. The female carries the offspring during prenatal development, nurses the baby after birth, and then cares for the child throughout the juvenile stage. Since she must spend more time and effort raising the child, the female is typically known as the higher investing sex. As the higher investing sex, the cost of reproduction is high, both biologically and socially. Due to the time and effort reproduction entails, females must be very selective about with whom they choose to reproduce. Ideally, females will only choose those males who demonstrate the highest genetic quality, the highest likelihood of staying with her and her children, and the most resources as reproductive partners. For many females, it is more beneficial to share a male with other women, if he has enough resources (biological or social) to provide for all of them, than to reproduce with a male who has no resources. This leads to the mating style of polygyny, which is the most common mating style throughout the world. This tendency for women to look for signs of commitment and ability to provide also leads to human males possessing traits that contribute to reproductive value rather than mere ability to survive.

In general, or on average, males are not biologically required to invest as much time and energy into offspring. They do not carry the offspring during the prenatal period, they do not nurse the baby after birth, and they are more likely to abandon offspring than are females. Due to these differences in biological demand, males are known as the lower investing sex. In many cultures, fathers do heavily invest in their offspring. Many fathers play an instrumental role in directly raising children, but even with the modern emphasis on marriage and dual parenting, there are still many more single mothers than single fathers, and males still do not bear the onus of pregnancy and the risk of childbirth.

Due to the tendency for males to provide less investment in offspring, being chosen as a partner frequently emerges as more instrumental than finding the highest quality mate. A man who can reproduce has successfully passed his traits to the next generation, regardless of how much he invests in the offspring. If he has several offspring with several women, he is even more biologically successful. This explains why males are more likely to be interested in sex, more likely to seek out sexual relationships, more likely to have more sexual partners, and less likely to be concerned with sustaining long-term relationships, on average. The biological emphasis for males is swayed toward quantity over quality, and traits that make them more attractive to a multitude of partners may be more beneficial than traits that allow them to live a longer life.

In summary, the role individuals play in the parenting process shapes what traits they find most attractive in a reproductive partner. For humans, in most cultures, males are the lower investing sex, biologically and socially. Thus, it benefits them to be attracted to many partners and have many offspring to pass their genetic material on to the next generation. Females, as the higher investing sex, are attracted to partners who will invest, protect, and help parent, which will make the children more likely to survive. These differences in parental investment explain many of the behavioral differences between men and women and many of the differences in traits of attraction.

THE UNIVERSAL ROLE OF SEXUAL SELECTION AND PARENTAL INVESTMENT IN ATTRACTION

Due to this universal gender difference in parental investment, researchers predicted that there should be differences in what each gender finds attractive and that these differences should be consistent throughout the world and over time. Since men and women play differing roles in the reproductive process, they have been subject to different evolutionary pressures. Women, in all cultures, bear the onus of pregnancy and childbirth, and thus, they should have similar requirements and desires when seeking a partner, and men, in all cultures, should have similar requirements when assessing the attractiveness of the opposite sex.

David Buss and colleagues from the University of Texas at Austin presented foundational research on how attractiveness is rated around the world. In their extensive International Mate Selection Project, Dr. Buss and colleagues gathered information from over 10,000 participants from 37 different cultures over the course of five years. Participants spanned six continents and five islands, and were made up of individuals from different religions, ethnic groups, races, political systems, economic systems, and mating systems (i.e., monogamous and polygamous groups were represented). As expected, culture did have an impact on which traits are considered most attractive. However, gender also emerged as a determining factor. That is, men, from all cultures, rated particular traits as more attractive than did women, and women, from all cultures, rated some traits as more attractive than did men. Therefore, some characteristics of attraction surpass cultural influences. Differences also emerged depending on what type of relationship was being considered. Platonic, short-term romantic, and long-term romantic relationships elicited differences in ideal traits. The specific findings from this study will be discussed throughout this chapter.

THE ROLE OF GENDER IN ATTRACTION AND RELATIONSHIPS

When examining relationships from a cross-cultural perspective, many fundamental differences emerge, such as differences in mating style, arranged marriages, and attitudes toward sex. For example, in the United States, most view marriages as monogamous, romantic affairs. We have legal sanctions against cheating, and sexual relations outside of the marriage are often a cause for divorce. Deviant groups portrayed on television shows such as *Sister Wives* are watched with a sense of awed or shocked fascination. However, although not common in the United States and Europe, polygamy is common in many other parts of the world. More than 40 countries in Asia and Africa still practice polygamy, and, in those countries, over 50 percent of women have co-wives. In other countries, practicality and politics surpass romance. For example, arranged marriages are still common in countries such as India, Pakistan, Japan, China, and Israel. Finally, some countries promote sexual exploration before marriage, others discourage it, and others strictly prohibit men and women from even being alone together until after wedlock. These cultural differences influence what individuals find attractive in a partner and explain some of the differences in the ratings of attractiveness throughout the world.

Culture, therefore, plays a role in attraction and preferences may shift over time. However, Buss demonstrated that some traits surpass cultural influences and are tied to gender rather than to culture. The next part of this chapter will focus on these universal and cross-generational gender differences.

A young newlywed Pakistani couple has just returned from their wedding ceremony. They were brought together in an arranged marriage as is common throughout many regions of Asia. Such marriages constitute an alliance of the individuals as well as their families. (Bernard Weil/Getty Images)

Traits Attractive to Women throughout the World

Many traits emerged as important to both men and women when selecting a mate or rating others for level of attractiveness. Both men and women report high standards, particularly when seeking a long-term partner. Men and women from all of the cultures identified that characteristics such as kindness, compassion, intelligence, dependability, similar values, similar humor, and good health were attractive in a potential partner. These characteristics were equally rated by males and by females in all of the groups and were considered to be important, especially for long-term partners. These qualities ensure compatibility and enjoyment when spending a lot of time with another individual. However, many differences between men and women emerged as well. These differences are not surprising, given the differences in sexual selection pressures discussed in the previous section.

In Buss's study, women, across all cultures, were significantly more likely than men to rate economic and physical resources as well as a willingness to share resources as significant factors when rating the level of attractiveness of a potential partner, and these preferences hold up over time. When asked to rate for the level of attractiveness, these personal qualities were actually rated as more important than any physical trait. Specifically, women reported that men who demonstrate signs of economic resources, financial stability, and high social status are more attractive than less financially successful men. These traits would ensure that the man has the resources to provide for the woman and their offspring, making it more likely that the offspring would survive to reproduce. In a later study, when women were asked to rate male photographs for the level of attractiveness, women provided significantly higher ratings of attractiveness for men when they were reported as being financially successful.

Younger males have not likely had the time or experience to achieve high social status or build resources, and so, women tend to be attracted to older, more successful men. Women, on average, prefer partners who are three to four years older than themselves, and who demonstrate ambition, industriousness, dependability, stability, and signs of willingness to share resources. If a woman is attracted to a younger male, he is likely one who displays enhanced signs of ambition and intelligence and who is highly likely to be successful. Laziness and lack of ambition are not likely to turn a woman's head.

Modern Women Provide for Themselves

Women's preferences stem back to when women needed a partner to provide resources for her and her offspring to survive. In earlier eras, without modern medical care or modern birth control, women spent a great deal of their adulthoods pregnant, nursing, and caring for young children. Without a partner to protect and to provide resources, children were less likely to survive. Thus, those women who found higher status, larger, more athletic, and committed males more attractive are the women whose children survived. All of these traits are indicators of good health and good genes as well as increased ability to protect offspring.

As men and women achieve more equality in the workforce, one might expect that women would no longer value traits such as economic and physical resources. In the modern era, many women do not need a partner to provide resources, and researchers wondered if female preferences would shift and become more similar to a successful male. Surprisingly, research shows that even a woman with resources of her own still rates these traits highly in a partner, maybe even more highly than less financially successful women. As a female's own wealth, power, and status increase, she has the luxury of being even more selective and demanding of a quality male. Studies find that women typically choose a mate who is at least on par with her level of socioeconomic status (SES), so as her SES increases, her pool of potential partners actually shrinks.

The opposite is true for men. Men have historically had control of the wealth and power, so men do not need to seek a woman who can provide those things. Instead, men must find a partner who can have healthy children to continue his genetic line. Thus, as a man's SES increases, his number of potential partners also increases because he can now also attract higher status and younger females who are looking for someone who can provide. Thus, men with higher status likely partner with younger, healthier, and more fertile women. This explains some relationships between wealthy older men and young beautiful women. Each benefit from securing ideal mates, even though those ideals vary by gender.

At the other end of the spectrum, women who lack their own economic or physical resources should be even more attracted to wealthier, higher status men. The desire for a male who can provide resources is so great that it explains why in countries where the distribution of wealth is highly variable, women may choose to share a wealthy man with other women rather than to have a poor man all to herself. In polygamous societies, it is not uncommon for the wealthiest men to garner the attention of many of the women while the poorest men receive no attention at all.

In order to examine actual behavior and not just self-report information about attractiveness, researchers analyzed personal ads to conclude what traits were most likely to prompt an interested response. Research reveals that women tended to respond to personal ads of taller men, who were described as three to four years older than themselves, who reported being successful, and who showed signs of being nurturing. Other physical traits did not significantly factor in. Wealthier, more successful women were the most selective and had the highest standards when selecting a long-term mate. Since resources were not a factor, these women had the luxury of selecting partners who demonstrated superior health, highest attractiveness, and most appealing personal qualities (including economic resources).

Traits Attractive to Males throughout the World

In Buss's International Mate Selection Project, men tended to be less complicated in their preferences than women. Men, across all cultures, were significantly more likely than women to identify physical traits as most important in a partner. In all 37 cultures studied, males rated physical indicators of youth and health as the most attractive in females. Traits such as full lips, smooth skin and clear eyes, shiny hair, long legs, good muscle tone, high energy, low waist-to-hip ratio (WHR), well-distributed body fat, high cheekbones, and symmetrical features all contributed to the ratings of attractiveness.

Evolution's Impact on Modern Attraction 105

From an evolutionary perspective, women's mate value is measured by her reproductive capability. Since fertility cannot be directly seen, it must be inferred. The biggest clues are age, physical features (age, WHR, symmetry), behavioral features (energy levels), and chastity (men do not want to father another man's child!). In 62 percent of the cultures studied, men preferred women without previous sexual experience. There were no cultures where women requested that of their male partners.

When analyzing the traits that males preferred, it is notable that anything that interferes with fertility is found to decrease the ratings of attractiveness. Obesity, malnutrition, pregnancy, and the physical effects of menopause all directly influence WHR. The ideal WHR as rated by males is 0.70. As discussed earlier, for every 10 percent rise, there is a 30 percent decrease in the likelihood of a successful pregnancy (regardless of age or overall weight). Therefore, males attracted to women with a low WHR were more likely to have more offspring, passing that natural preference to modern males.

Of course, men are also attracted to intelligence, kindness, and sense of humor in a partner, but the physical features were found to be a much greater driving force for men than for women. Men, in direct reciprocity with women, were more attracted to women who were younger than themselves. On average, the ideal wife was rated to be two and a half years younger, impacting her likely levels of fertility.

It would be easy to write men off as superficial since they are so focused on physical qualities, but their choices and preferences make evolutionary sense. A younger, more attractive woman is statistically more likely to bear more children who are healthier and more apt to survive to reproduce themselves. So, past males who selected young, fit women were more likely to have more offspring who survived and reproduced as well. If this preference is genetic, then those multitudes of offspring carry that preference with them through evolutionary time. Furthermore, there is an exception to this rule of attraction to younger women. Since women are typically the most fertile in their twenties, evolution would predict that this age group should be the most attractive to males, regardless of the male's age. When analyzing this, researchers found that younger males (between 16 and 20) actually found these older, more fertile women more attractive. This is a clear exception to the rule that men are attractive to younger women. It may be more accurate to say that men are attracted to more fertile women. Of course, it is not reasonable that all males can reproduce with the same few females, so concessions must be made and the best possible partner selected.

Throughout the cultures studied, there were some interesting cultural differences in men's preferences. Although a low WHR was considered attractive by men across all cultures studied, preferred overall body size (body mass index or BMI) varied. Upon examination, however, this difference is explainable given the aspects of the societies. Economic systems and economic stability of the society impacted the preferences of the individuals within the culture. The most attractive body size for a woman was tied to social standing. As discussed earlier in this book, in poor societies, heavier women observably have access to more food, contributing to their plumper figures and indicating higher social status and greater attractiveness. In rich societies where food acquisition is not an issue, thinner women are preferred and the lower status individuals are more likely to be overweight due to the poor quality of cheaper food. So body mass does link to attractiveness and beauty but the ideal BMI varies by culture. The ideal WHR stayed consistent across cultures, however. Larger or smaller women can still have a small waist in comparison to their hips, creating the feminine hourglass figure. This WHR has ties to reproductive and overall health and seems to be a more important factor in attractiveness than overall body weight.

A Misconception between the Sexes

It is interesting to note that in 26 of the cultures studied, particularly the industrialized societies, males preferred heavier females than the females themselves predicted. When asked to identify the ideal female silhouette, women chose a slimmer figure as what they thought men would find most attractive than the men actually chose. Thus, the preference for excessive thinness for women seen in cultures such as the United States is not necessarily a by-product of male interest but a by-product of what women think men want.

THE ROLE OF CONTEXT IN ATTRACTION

Individuals tend to pay more attention to a partner whom they find physically attractive. Although the evolutionary basis of attraction is for reproduction, humans engage in relationships for a host of reasons beyond procreation. Research shows that when rating attractiveness, the context matters. Research shows stark differences in choices and behaviors when we are pursuing different types of relationships. For example, individual judgments about attractiveness vary predictably depending on the type of relationship one is seeking (friend vs. partner), length of desired relationship (short-term or long-term), and environmental circumstances (economic climate or incidence of disease).

Friendships and Cooperative Alliances

For friendships or cooperative relationships, research reveals that individuals tend to be attracted to familiarity and clues of genetic similarity. People tend to trust and be more trustworthy when working within family or kin groups, making these individuals better cooperative partners. For reproductive purposes, however, high genetic similarity is not as beneficial. On those occasions, individuals tend to be attracted to those who are less genetically similar. In a study where participants were asked to judge a photograph for level of attractiveness for a cooperative alliance and for level of attractiveness as a sexual partner, participants rated images that looked more similar to themselves as more attractive for cooperative partners but these same images were not rated highly as potential sexual partners.

Short-Term Sexual Relationships

Short-term sexual relationships are those in which individuals engage in sexual activity with a partner without any commitment to a long-term relationship. When considering short-term relationships, ratings of attractiveness differ from when considering friendships or long-term sexual relationships for both men and women. Individuals engaging in short-term affairs may hold the hope that a long-term connection will develop, or they may be just enjoying sexual intimacy with no expectation of a future relationship. Short-term relationships are more common during college, on vacations, or directly after a breakup of a long-term relationship.

Men endorse being more open to short-term sexual relationships than women and are more likely to rate available women as more attractive, regardless of their actual physical traits. This makes sense from an evolutionary perspective because men stand to gain more by engaging in short-term affairs than do women. Men could potentially pass on their genes to offspring without losing many resources. Women, however, even with short-term relationships, bear the risk of pregnancy, birth, breastfeeding, and child care, regardless of the lack of long-term commitment. Thus, men's standards for a short-term mate tend to be relatively low, at least compared to their long-term standards for a partner. Women's standards for a short-term mate, however, tend to be similar to or even stricter than those for a long-term mate. If a woman bears the risk of pregnancy from a short-term affair, it had better be worth it. For many men, the ideal outcome of a short-term relationship may be sex. For many women, the ideal outcome may be for it to develop into a long-term relationship. Thus, women want to make sure the partner is also someone they would enjoy for a long-term relationship, whereas men just need someone who is interested in sex.

Despite the risks, we obviously live in a culture where both men and women engage in promiscuous behavior. However, the motivation behind such behavior differs by gender. In the United States, birth control and modern medical care may have eliminated some of the concern for women, but women are still biologically predisposed to be selective, even when just seeking pleasure and excitement. For short-term relationships, most individuals seek exciting, physically stimulating interactions and place less importance on personality qualities. Women engaging in short-term romantic affairs rate masculine, powerful, and virile men as the most attractive and the objects of their extramarital affairs tend to be better looking, be wealthier, and have better genetic quality than their long-term partners. Benefits of this behavior could be to conceive a child with better genes or potentially solidifying more resources for all future offspring. Risks include losing the support of her long-term partner if she is discovered or raising a child without a partner. Based on these costs and benefits, it is less likely for a female to cheat on her long-term partner than it may be for males.

Given the same parameters, for males, we see a sharp decline in standards for a partner when considering a short-term affair rather than a long-term relationship. A short-term affair gives the benefits of sex and possible reproduction without bearing any of the risks of commitment or support for the resulting offspring. Although this logic is likely not present in the consciousness of the male, those who engaged in such behavior in our evolutionary past were more likely to spread their genes with more partners (greater variability) without having to commit and support only one partner. Since their genes were spread more copiously, this tendency to seek out short-term affairs was passed on as well. Thus, men engaging in short-term relationships rate willing sexual partners as the most attractive, and place less emphasis on physical or psychological characteristics for temporary partners.

In accordance with this logic, in Buss's study, when asked to rate potential partners for short-term relationships, men's standards dropped dramatically. Women, whom they may not have even considered as a partner, suddenly seem like viable choice for a one-night stand. For men, engaging in sexual intercourse confers all of the potential benefits of having sex, with fewer of the risks that women face. The only risk for the male is that he risks garnering a reputation that may make him less desirable to a quality female who is looking for a long-term relationship. Since females may need the support when caring for young, they are likely to prefer males who do not engage in promiscuous behavior and who show signs of commitment and monogamy. Further risks for the male may also be present in our current modern legal and medical system if he is identified as the father and forced to support the resulting children, but these are modern concerns that likely do not inform our evolutionary instincts.

These results remained the same even when examining men and women across cultures. Since women bear the brunt of reproductive investment, their standards, even for a short-term mate, remained high across all of the cultures studied. Although reproduction might not be the goal of a one-time affair, women, particularly women in our evolutionary past, still bore the risk of conception. Once pregnant, they may have been less able to find food and shelter, and likely had limited access to health care, particularly without a partner to share the load and his resources. So, women engaging in short-term affairs bear more risk to future survival than their male counterparts. Accordingly, Buss's research demonstrates that women's standards and ratings of attractiveness stay high when considering a short-term partner, perhaps even becoming more selective, since she is bearing the risk of pregnancy without commitment and support.

An examination of the impacts of these differential psychological mechanisms at play between men and women for short-term relationship scenarios was carried out in a college bar setting. Researchers asked bar patrons to rate the attractiveness of the other people in the bar throughout an evening. Preliminary results showed that men's ratings of women in the bar setting radically increased as closing time drew near. For example, when asked to rate the level of attractiveness of others in the bar early in the evening, men may have rated a particular woman as a 5 or 6 on a scale of 1–10. As closing time nears, this rating jumped to a 9 or a 10. As the window for a connection drew to a close, men were much more willing to compromise their standards to ensure a date for the evening. Even more interesting, these results persisted even for those who had not consumed alcohol over the night.

Similar findings emerged for women. As closing time drew near, women's ratings of the attractiveness of others also increased. However, women's ratings changed much less significantly. For example women's ratings of men in the bar changed from a 6 to a 6.5, demonstrating that women's standards remain high even if the opportunity is slipping away. Thus, for short-term affairs, or possible one-night stands resulting from a night out, men in actual bar settings were much more likely to alter their standards but women maintained their standards, likely due to the risk associated with uncommitted sexual activity.

So, expectedly, for short-term relationships, we do see bigger differences in attractiveness ratings between the sexes since women are risking more than men. For a one-night stand, men can relax their standards because they are not investing any resources, aside from a few million sperm. Men produce about 12 million sperm an hour so they do not have to be as selective about with whom they share them. Their reproductive success does not rely on their own ability to gestate, lactate, nurture, protect, and feed, but only on their ability to spread their seed. Women, however, constantly carry the risk of pregnancy and, without the promise of a stable partner to help, their risk increases. Furthermore, women only have approximately 400 ova over their lifetimes so they must be selective when choosing a partner. This would suggest that, for a woman, a short-term coupling is a risky affair, so her standards should remain high.

Dr. Michael W. Wiederman and Dr. Stephanie L. Dubois from Ball State University also examined sex differences in mate preferences for long-term versus short-term mates. In Dr. Wiederman and Dr. Dubois's study, men were more likely to endorse short-term relationships as desirable. Men emphasized physical attractiveness as an important quality for a short-term partner and were not as concerned with psychological or personality traits. Even annoying traits were tolerable for short-term relationships as long as the individual was physically attractive. Women were less interested in short-term relationships and were more likely to demand excellent personality traits as well as physical traits if considering short-term relationships. Thus, similar to findings from Buss's study, women's standards became stricter when considering the short-term, while men's became less demanding.

The fact remains that many women, particularly modern women, do engage in short-term affairs. One reason women may seek short-term relationships is due to hormones associated with ovulation. Women tend to be more sexually interested during the fertile phase of their cycles, and thus may be biologically driven to find a romantic partner at certain points during the month. During this most fertile phase, women are attracted to men with greater levels of testosterone who tend to be more risk taking, more dominant, and, consequently, less committed. During this point in their hormonal cycle, women are more likely to rate masculine men as having the potential to be good fathers. During the phase of lower fertility, women rated these men as being less faithful, more likely to cheat, and less likely to be a good father. This shows that hormones can alter the perception of mate quality and may be one reason that women engage in short-term affairs.

Long-Term Sexual Relationships

For long-term relationships, since both partners contribute and will be required to spend a lot of time with one another, researchers predictably expect that both men and women will maintain high standards. Although men and women may be looking for different things, they are both choosing to invest in a long-term relationship and so it serves them to be picky about whom they choose for a partner.

In an array of research studies, subtle and not so subtle differences were found when examining preferences for a long-term mate. For example, when seeking a long-term partner, women prefer less masculinized faces than they do for short-term partners. More masculine faces likely represent men with higher levels of testosterone and higher reproductive potential. Although this would seem to be a benefit because it increases the likelihood of successful procreation, it also increases the likelihood of risk taking, cheating behavior, and lack of investment. For long-term partners, masculinity becomes less important, and women instead prefer more familiar, compatible, and committed partners who are willing to invest resources and who will be emotionally, physically, and financially supportive.

As discussed at the beginning of this chapter, men and women have many similar preferences when rating potential long-term mates. Although there are some gender differences for traits that signal fertility and commitment, both men and women seek physically, psychologically, and emotionally fulfilling long-term connections. For women, an ideal partner would display traits of commitment and the ability to provide, but also be supportive and enjoyable to spend time with. For men, long-term partners should not only demonstrate signs of fertility but also be attractive, loving, and well matched. Making a commitment to spend a future with another person raises the stakes beyond the physical and males' standards for a long-term relationship predictably rival those of females'.

THE IMPACT OF ENVIRONMENTAL CIRCUMSTANCES AND THEORIES OF ATTRACTION

Environmental circumstances within a society can influence the ratings of attractiveness and relationship styles. For example, in societies where wealth is not evenly distributed, polygyny is likely to develop. Polygyny is the mating style where a man has multiple long-term female partners. In an environment where few men control most of the wealth, if a man has enough wealth to provide for multiple wives and offspring, a mutually beneficial trade-off emerges. Women in these circumstances rate ability and likelihood to provide as even more attractive than physical traits or monogamous commitment. In circumstances where the environment is particularly harsh and it is difficult to survive or to care for offspring, polyandry may emerge. Polyandry is the mating style where a woman has multiple long-term male partners. In a harsh environment, it may take more than one male to provide for an offspring, so men combine their efforts to raise a family. In these circumstances, the men are usually brothers, so the offspring are genetically related to both men, even if only one actually fathered each child.

In environments where there are high rates of disease, preferences for genetically diverse partners are accentuated, leading to more promiscuous behaviors for both genders. Mating with genetically diverse partners means that offspring will be more genetically varied and more likely to survive in changing environments. In such environments, men and women rate those who have more diverse traits as more attractive and tend to mate with multiple partners, which ensures greater genetic variability in the offspring.

The mating style of a society is also predictable based on the number of males compared to the number of females, or the sex ratio of the group. If there is an imbalance, the scarcer sex has the luxury of enforcing its own desires and the sex with more representation must bend to the other sex's will to have the opportunity to reproduce, as discussed in the next section. Finally, social effects including costs and benefits for engaging in relationships influence how relationships develop and are maintained.

Sex Ratio Theory

The sex ratio theory essentially refers to the ratio of how many reproductively viable males live in a given society as compared to the number of reproductively viable females. A high sex ratio refers to the circumstance where men outnumber women, and a low sex ratio refers to a society where women outnumber men. If one knows the ratio of men to women in a given population, predictions can be made about the mating style of the group.

In a population with a high sex ratio, there are few fertile women. Thus, the men must compete for the women's attention and women can pick with whom they would like to engage. In this circumstance, the man who aligns with a woman's preference is most likely to be chosen and the most reproductively successful. Since women benefit most from monogamous relationships with prosperous males, those men who demonstrate resources and a willingness to commit will be the most likely to be rated as attractive. This resulting society would most likely be monogamous, and men would devoutly invest in their partners due to scarcity of other options.

In a population with a low sex ratio, there are few reproductively viable men. In such a society, the women must compete for the opportunity to reproduce and men would have the luxury of choosing partners who have the traits they desire. In this circumstance, the woman who aligns with the man's preferences is most likely to be rated as most attractive. Therefore, women who demonstrate youth, attractiveness, and an interest in casual sexual intercourse would have the most opportunities. Most likely, in this society, promiscuity or polygyny will emerge. Women may opt to seize the opportunity to mate with a prosperous male and may be willing to share a wealthy man to ensure resources and reproductive success.

Sociosexual Orientation and the Strategic Pluralism Theory

The sex ratio within a given society affects the overall sociosexual orientation of that group. Sociosexual orientation refers to the level of promiscuity or the likelihood that individuals will participate in sexual experiences outside of long-term relationships. Individuals with an unrestricted sociosexual style are more likely to engage in sex without emotional closeness or commitment. A restricted style describes someone who is not likely to engage in sex outside of a committed relationship. Sociosexual orientation tends to vary predictably by gender and culture. Men, on average, have a more unrestricted style and women have a more restricted approach to sex. In cultures where men outnumber women (high sex ratio), men must conform to the women's style if they want to secure a partner, so the culture as a whole endorses traits that align with more restricted, less promiscuous, and more monogamous behavior. In cultures where women outnumber men (low sex ratio), women are more likely to conform to the desires of the men in order to secure a partner, and thus women become less restricted and more promiscuous.

Mating behavior, although genetically driven, is obviously at the mercy of many environmental influences. The strategic pluralism theory suggests that humans have multiple (plural) strategies that they may pursue to successfully reproduce. Which strategy is the most beneficial depends on the given environment. Thus, genetic behaviors may be triggered by environmental circumstances. Sociosexual orientation varies predictably given the environmental conditions and can be triggered by sex ratio, luxury, disease, or other hardship.

Social Exchange Theory

There is no question that our genetics, environment, and evolutionary history influence our natural preferences and partner selection. However, from a social psychological point of view, there are other influences that emerge as important when selecting a partner. Furthermore, these social influences tie into our genetic tendencies. The social exchange theory suggests that we are attracted to those who provide us with high rewards at a low cost. These rewards may include things such as personal enjoyment, positive affect (e.g., happiness, satisfaction), emotional or physical support, or reproductive success. Costs could include things such as negative interactions, stress, draining of resources, or poor quality offspring. In humans, women bear a higher cost when bearing offspring. Thus, for women, the rewards provided by a partner must be quite high to alleviate the costs accrued, especially if engaging in a short-term fling. For men, offspring can be less costly, and thus, their costs accrued can be very low, leading them to have lower standards when seeking a short-term mate.

As predicted by David Buss, from an evolutionary perspective, men should predominantly prefer promiscuous sexual encounters because each encounter has the potential to pass on their genes. In this circumstance, there is high reward (their genes are passed to the next generation) and low costs (if they are not committed to the relationship, they only waste a few minutes and a few million sperm). Even if their lack of investment means that not all of the children they create survive, they can still have evolutionary success as long as some of them do.

Women, however, have much greater investment (biologically and culturally) should a child be conceived, and can only have one child every nine months or so; so women should prefer monogamous sexual encounters with men who have resources and demonstrate a willingness to provide. Women have high reward (their genes are passed to the next generation) but they also have high costs (they must carry, deliver, and care for the child). Thus, men and women have likely evolved to prefer many different qualities in mates, creating a tug of war in the relationship process. Based on an evolutionary drive, men should prefer fertility, typically displayed through signs of youth and beauty, and women should prefer signs of status, commitment, and wealth.

The social exchange theory predicts that social relationships involve exchanges. In this context, women will exchange their youth and attractiveness for men with status and resources. Thus, women with the highest ratings of attractiveness should have mates with high status and high levels of commitment, and males with the highest status should have mates with the highest levels of attractiveness or several mates with moderate levels of attractiveness.

Dr. Eugene W. Mathes and Dr. Ginger Kozak from West Illinois University sought to explore how this prediction emerges in real relationships. They worked with 56 undergraduate women who were in long-term relationships (a month or more). Participants filled out surveys regarding the status, motivation, commitment, and personality of their partners and then they were each rated on their level of physical attractiveness. Results showed that the women who were rated among the top third most attractive women in the study had partners with significantly greater resource potential. The partners of these women were rated as more motivated, ambitious, financially stable, and hard-working and as possessing more positive personality traits such as being more kind, stable, intelligent, and mature than the partners of the moderately attractive or unattractive women. The highly attractive women also reported that their partners' levels of commitment were higher than the other two groups.

Dr. Mathes and Dr. Kozak did not find significant differences in the level of sexual experience between the highly attractive and less attractive groups. This is not particularly surprising because more attractive women are more sought after for sexual partners, but are also pickier. Less attractive women are less sought after, but are less picky, so the average number of partners for each balances out. The more attractive women may have had more sexual experience in the context of long-term relationships and the less attractive women may have had more short-term experience, but this was not tested.

As illustrated in Chapter 2, attractiveness is highly correlated with a whole host of other benefits. Due to males' biological tendencies toward more promiscuous drives, women's level of sexual experience is likely limited only by their own desires and demands. More attractive women likely have higher self-esteem, are more assertive, and have more relationship and business opportunities. Dr. Mathes and Dr. Kozak conclude that more attractive women use their level of attractiveness to acquire mates who have the highest ability to provide for themselves and potential offspring.

Most women have the opportunity to reproduce if they so choose. The likelihood of reproduction for men, however, is more variable. Wealthy, charismatic, competent men can have sexual relationships with many women, while their counterparts may not get to have sex with any. Thus, male traits are more susceptible to evolutionary pressures because only those with the best qualities have the opportunity to pass on their genes. Successful men tend to be bigger, stronger, and more violent toward their competition. They tend to take more risks, expend more time and energy on reproductive pursuits, and may pursue a greater assortment of partners. They tend to take longer to mature, die earlier, and are more aggressive. Women select men based on these traits. A male with power, strength, and courage is more attractive than one who merely has appealing physical traits.

CONCLUSION

In our evolutionary history, we did not have modern medical interventions to alter our natural features. Thus, physical traits were more honest and informative than they are today. Having information about immune system functioning, genetic quality, and social value was highly beneficial when we relied on partners for survival and reproductive success. The importance and informative value of evaluating physical traits thus led some individuals to be more reproductively successful than others. Those who were more reproductively successful had the advantage of passing on their genes (along with their natural preferences) down through evolutionary time and into the modern generation. Thus, even though we may be fooled by physical traits in today's society that includes cosmetics, cosmetic surgery, and Photoshop, we are still programmed to rely on the cues that proved to be successful in earlier generations.

This evolutionary view of attractiveness explains why some traits are cross-culturally more attractive than others, providing a common viewpoint for a base level of attractiveness. Those with traits that indicate greater genetic quality, better physical health, greater earning potential, and higher commitment to offspring are maintained in the population through the process of sexual selection. This means that some individuals are more likely to be selected as mating partners and thus more likely to pass on these traits to offspring. These traits may or may not provide an actual advantage for survival, but they do provide an increased potential for having offspring.

Research demonstrates that culture, gender, and environmental context all help explain attraction and the underlying science behind human attractiveness. Evolutionary concepts such as natural selection, sexual selection, and parental investment help explain why some traits are more attractive than others, particularly from the perspective of gender differences. The sex ratio theory explains why particular relationship styles have emerged within specific cultures, which determines the overall sociosexual orientation of the group. Furthermore, the strategic pluralism theory and social exchange theory help explain why ratings of attractive traits differ given the type and length of relationship sought, environmental circumstances of the society, and the importance of social exchange within relationships. In summary, attractive traits tend to be those that increase quantity and quality of offspring, increase interpersonal cohesiveness, and increase individuals' abilities to achieve personal and professional goals.

6
The Impact of Attractiveness on Behavior and Relationship Satisfaction

In Chapter 2, the benefits of being attractive were explored from a social and psychological perspective. This research illustrated that there are multiple benefits of being attractive including garnering more attention, more opportunities, more money, and more praise, and developing higher self-esteem. Attractiveness is also a contributing factor to many aspects of relationship initiation and maintenance. Thus, this chapter focuses on the impact of attractiveness and attraction on behavior, specifically with regard to romantic relationships.

Research has demonstrated that one's level of attractiveness can influence relationship initiation, longevity, quality, and satisfaction, but that not all the effects are positive. As will be explored in this chapter, relationship satisfaction is influenced by intimacy, level of commitment, and personality, all of which are influenced by attractiveness. However, high levels of attractiveness tend to correlate with more flirtatious behavior, increased mate-poaching attempts (attempts by others to steal one's partner), and more difficulty sustaining relationships. Thus, jealousy and divorce tend to be correlated with higher levels of attractiveness. Men and women demonstrate what they find attractive based on the content of their sexual fantasies, flirtations, and pornography use, so these topics will also be examined. Understanding the relationship between attractiveness and behavior can explain many aspects and difficulties of relationships. Although having a high level of attractiveness can facilitate the initiation of relationships, attractiveness can also instigate feelings of jealousy from a partner and mate-poaching attempts by other interested parties. Thus, a high level of attractiveness can actually complicate and damage long-term relationships.

INITIATION OF RELATIONSHIPS

Attractive individuals tend to be more sought after for many types of relationships including friendships, business partnerships, and romantic relationships. William G. Graziano and Jennifer Weisho Bruce examined why attractive individuals are pursued more for romantic relationships. They found that the reasons others seek out attractive individuals vary. When specifically viewing an attractive individual, the ventral occipital region (an area in the visual cortex at the back of the brain) is activated. Thus, there seems to be built in hardware that facilitates the recognition of attractiveness, making attractive individuals more likely to capture our attention. Viewing an attractive individual also likely activates culturally acquired stereotypes about personality traits and likely future behaviors. These stereotypes include the assumption that an attractive person will be more socially skilled and friendly, which are both traits that are engaging to others. So, we may be biologically *and* socially attracted to a beautiful face.

We are biologically predisposed to notice attractive individuals, and our expectations of personal qualities that tend to accompany such traits increase our interest and the likelihood that we will pursue a relationship, simply based on the expectations the physical characteristics create. Dr. Graziano and Dr. Bruce also found that the effects are reciprocal. This means that someone who is known to be socially skilled and friendly is actually perceived as more physically attractive simply due to his/her behaviors. Thus, not only does physical appearance influence our ratings of other qualities, but also knowing about other qualities can actually influence our perceptions of the physical appeal of others. Furthermore, research shows that both genders tend to be more romantically attractive to and attracted to others who demonstrate a reciprocal interest. Eye contact and directed smiling behavior increase level of interest for the rater, creating a positive feedback loop. So, physical attractiveness, signs of good social skills, and signals of reciprocal interest all facilitate initiation of new relationships.

RELATIONSHIP LONGEVITY AND QUALITY

Although attractive individuals are more sought after for relationships, subsequent research has demonstrated that attractive people may have more difficulty sustaining those relationships. Males, in particular, rate physical attractiveness as highly important for romantic relationships, and if there are several attractive alternatives, it may be more difficult for men to make a long-term commitment. In this sense, some research has suggested that physical attractiveness may not always be beneficial for relationship satisfaction or longevity. High levels of attractiveness have been correlated with higher temptations to cheat on a partner, more opportunities to cheat, and more cheating behavior overall. Although physical appeal tends to be more important to men, the problem of sustaining committed relationships holds true from both genders. More symmetrical men admit to more affairs and more short-term partners, and research shows that they have higher divorce rates than their less attractive counterparts. However, women with a more ideal waist-to-hip ratio (WHR) also report more extra-relationship affairs than women with higher WHR measurements. Highly attractive individuals of both genders report being hit on more frequently, even when others know that they are in committed relationships. Furthermore, these mate-poaching attempts may be from other highly successfully and highly attractive individuals, making them more difficult to resist.

Christine Ma-Kellams, Margaret C. Wang, and Hannah Cardiel from Harvard University examined the impact of rated levels of physical attractiveness and self-perceived levels of physical attractiveness on relationship outcomes. These researchers had female participants rate the level of attractiveness of 238 men (using old high school yearbook photos), and then the researchers examined subsequent archival data to examine the correlation between the men's rated level of attractiveness and their rates of marriage and divorce over the following 30-year period. Not only were the males who were rated as more physically attractive more likely to be divorced, but the level of attractiveness was also significantly correlated with the length of marriage. More specifically, the researchers found that the more attractive the individual, the shorter the marriage. When this study was extended to include even more highly attractive males, such as actors, actresses, and celebrities, the results were replicated. The more attractive celebrities were married for shorter periods of time, on average.

Since length of marriage does not necessarily describe relationship quality, Dr. Ma-Kellams, Dr. Wang, and Dr. Cardiel subsequently expanded their research. They found that an interaction effect between relationship satisfaction and level of attractiveness contributed to the likelihood of ending a relationship or cheating on a partner. Specifically, attractive individuals who were not satisfied with their current relationships were more likely to rate extra-relationship possibilities as more attractive and to demonstrate more interest in others even while in a relationship. Attractive participants, who were happy in their current relationships, did not show this increased interest in extra-relationship partners. Furthermore, less attractive individuals and attractive individuals who were made to feel less attractive (manipulated by showing them pictures of highly attractive same-sex individuals) were less likely to be interested in extra-relationship possibilities regardless of the quality of their current relationships. Thus, relationship satisfaction is particularly important for relationship duration, particularly for attractive individuals.

Examination of this and several other studies that survey relationship satisfaction repeatedly reveals that high levels of physical attractiveness are correlated with briefer relationships, higher levels of disengagement in the relationship, and a greater likelihood of the relationship ending. Physically attractive individuals are more likely to get divorced and have shorter marriages overall. Attractive individuals are also more likely to seek alternatives if they are not satisfied with their current partners. Alternatively, less attractive individuals were more likely to have increased commitment to their partners, longer marriages, and less interest in alternatives even during times of low satisfaction. So, not only do others pursue highly attractive individuals but attractive individuals also show more interest in others, especially during periods of poor relationship satisfaction.

COMMITMENT AND JEALOUSY

Since jealousy tends to emerge in times of threat to a relationship (even perceived or imagined threat), or in times of threat to an individual's self-esteem within a relationship, researchers were interested in how attractiveness influences relationship stability. Arie Nadler and Iris Dotan from Tel-Aviv University examined the antecedents to jealousy in romantic relationships. They found that gender, degree of threat, and the level of attractiveness can all influence romantic jealousy. If a partner is highly attractive, the individual experiences more threat from other people who may be seeking to poach the partner from the relationship. If there is a rival, the rival's level of attractiveness can further increase the feeling of threat (and, hence, jealousy). The impact and source of the threat vary by gender. Males tend to feel more upset over the loss of status due to the threat, while women are more worried about losing the relationship. Thus, men are more jealous when a partner shows feelings of sexual attraction to another male who is of high status, which undermines the current male's status. The woman is more likely to be jealous when her partner shows commitment to another person, even devoid of sexual attraction, because committing to another threatens the relationship.

To further examine the impact of jealousy on relationships, Dr. Nadler and Dr. Dotan asked married adults to read stories about situations that might provoke a jealous response and then to discuss how the character in the story might feel or respond in the situation. These stories were about men or women who were involved in extramarital affairs. The characters differed by the level of attractiveness and status of the extramarital partner or the level of commitment to that extramarital relationship. Participants were asked questions about the protagonist's predicted emotions, physiological responses, behaviors, and attitudes toward the spouse.

Findings revealed that jealousy, unsurprisingly, was correlated with the level of the rival's attractiveness and how committed the spouse was to the extramarital relationship. Women expressed higher feelings of jealousy when the rival was attractive *and* the relationship was described as one of high commitment. Females rated lower levels of jealousy when the relationship was described as one of low commitment, regardless of the attractiveness of the rival. Thus, the quality of the relationship emerged as more important to their reactions than the physical features of the other woman. Women were more concerned if their partners show commitment to another regardless of the other's level of attractiveness. Attractiveness of other women is less threatening if their partner shows no commitment to the other.

Men expressed high levels of jealousy when the other man was described as attractive and successful. However, if this extramarital affair with a highly attractive rival was brief and noncommitted, then men rated being more attracted to their wives. If she was described as being committed to the rival, the male participants predicted that the husband would find her less attractive. For both men and women in the study, highly attractive and highly committed relationships with extramarital partners aroused the highest ratings of jealousy. A situation where a rival is highly attractive and a spouse is highly committed in the extramarital relationship would be the most threatening to the current relationship. Since the rival is attractive, it also threatens the individual's self-esteem as well as the current relationship. If the relationship is only temporary and noncommitted, the individual seems to feel less threat. This is also the case when the rival is described as of low status and unattractive. There is less threat to the self or to the relationship in these situations, thus eliciting lower levels of predicted jealousy and other traits of negative affect (e.g., sadness, disappointment, anger, frustration).

In these situations, women and men reported different strategies to deal with the relationship problems. Women tended to recommend confrontation or expression of negative emotions with their partner and men tended to endorse ending the relationship or avoiding the partner and would not admit to high levels of negative emotions. Women were seen to approach and try to protect the relationship by exposing their negative emotions in relation to their partner betraying their trust and men tended to cope with the threat by endorsing things like seeing other women or ending the marriage.

Pushing Back against Jealousy

Jealousy can be detrimental to a relationship or provide an indication that there is a problem in the relationship. Those relationships that demonstrate the highest quality (and least jealousy) tend to include relationships with couples who are highly committed to one another. In high-quality relationships, individuals engage in more maintenance behaviors like active communication about their feelings and engage in exchanges using positive affect, and they tend to more honestly discuss their feelings, particularly when threatened or when feeling jealous. Couples tend to play off one another and determine their own level of commitment based on their partner's level of commitment.

Because there may always be threat from other attractive individuals for established couples, Dr. John E. Lydon from McGill University and colleagues from across North America examined what it takes for couples to be secure in their relationships and be less concerned about other attractive people. They compared individuals who were not in committed relationships, individuals who were in committed relationships but not married or married but low on scales of commitment, and individuals who were married and highly committed to their partners. Under the guise of helping develop and test a dating application, in the first phase of the study, individuals were asked to create a biography, answer relationship questions, and submit a photograph.

In the second phase of the study, individuals thought they were assessing biographies of other participants in the study. At this stage, the level of threat was varied. In the moderate threat condition, the individuals were asked to rate another participant's dating profile (all participants rated the same opposite-sex photo of a highly attractive, single, and reportedly available confederate). In the high-threat condition, the individual was told that the person they were rating had already viewed the participant's profile and had expressed attraction to them after viewing their biography and photograph. In this condition, the subject arguably believed they could pursue a relationship with this other individual if they so chose. In the first condition with moderate threat (simply rating an attractive profile), both the low-commitment and high-commitment individuals rated the other individuals as attractive. However, in the high-threat condition, highly committed individuals rated the attractive other as less attractive. Thus, when in a highly committed relationship, individuals assess alternatives as less attractive than they would when not in a committed relationship. This shows that people in high-quality committed relationships actually skew their own perceptions to maintain relationships and protect relationship satisfaction. So, individuals in a highly committed relationship did not respond to the invitation from a highly attractive alternative but actually judged the alternative to be less attractive than they would have in other situations.

The Impact of Attractiveness on Behavior and Relationship Satisfaction 123

This research shows that commitment in a relationship is impacted by several factors, including the level of satisfaction within the current relationship, the number and extent of investments made, and the attractiveness of potential alternatives. Additional research shows that commitment is also impacted by the amount of stress in the environment, the quality of communication between the individuals, and the amount of conflict. If problems arise, there are several potential ways to address the issues. Individuals can end the relationship, talk about the problems, have faith the conflict with pass, or ignore the problems or the partner. Highly committed partners tend to do more talking and have more faith in their partners. Furthermore, level of commitment tends to increase if there are not readily available attractive alternatives.

Sternberg's Triarchic Theory of Love claims that successful long-term romantic relationships need passion, commitment, and intimacy. The element of passion requires physical attraction but this attraction alone is not enough to ensure a successful long-term relationship. Commitment and intimacy may require effort, while passion may be natural and outside of one's control. Thus, passionate attraction is likely the simplest piece of the equation when building a solid relationship. When relationships were analyzed, level of commitment emerged as more important than physical attraction for long-term relationship success. This commitment can be monetary in nature or include shared resources and materialistic possessions, but it is just as likely a by-product of the culmination of shared experiences, memories, friends, and future plans that would be lost if the relationship ended.

Conflict resolution and communication styles also emerged as highly predictive of relationship quality and level of commitment. The predictive value of these theories varied between genders. Men are more likely to maintain a committed relationship if they are satisfied with their partner and if the alternatives are poor. For women, investment emerged as the most important predictor of commitment. To increase commitment, both individuals need to be able to communicate, respect, appreciate, and understand one another. Commitment requires patience, kindness, and honesty and cannot be realized from one side alone. Both partners must choose to invest to ensure a high-quality and highly satisfying relationship to establish commitment and intimacy, as discussed in the next section.

THE EFFECTS OF INTIMACY AND PERSONALITY ON RELATIONSHIP SATISFACTION

As Sternberg presents, passion, commitment, and intimacy contribute to relationship quality. Passion revolves around physical appeal, and commitment involves the intention to sustain the relationship over the long-term. Intimacy is the third aspect of successful relationships and can vary over many different aspects, including emotional intimacy, intellectual intimacy, spiritual intimacy, and physical intimacy. While physical attraction may be helpful in beginning a relationship, intimacy may be more important to help sustain it. Emotional intimacy revolves around the support, affection, and understanding within a relationship. Intellectual intimacy provides a cognitive connection between the two people and allows them to feel comfortable sharing thoughts and ideas. Spiritual intimacy provides a common ground for faith and values and a sense of excitement when thinking about or being with the other person. And, physical intimacy, separate from sex, provides physical connection and the benefits of touch.

Similar to physical attractiveness, researchers demonstrate that intimacy directly effects relationship quality, longevity, and satisfaction. Furthermore, personality traits are likely to affect the levels of intimacy. Dr. Robert J. Taormina and Dr. Ivy K. M. Ho from the University of Macau in China specifically expected that personal characteristics could influence the level of intimacy in relationships, which would influence the level of relationship satisfaction. These researchers examined personality traits that include emotional intelligence, personality dimensions from the Big-Five model, as well as physical attractiveness, self-esteem, values, trust, locus of control, and gender to see how each affects intimacy. Specifically they found that emotional intelligence, or the ability to use and manage one's emotions in social interactions, contributes to overall relationship satisfaction. Those who are able to recognize, manage, and regulate their own emotions have stronger and more satisfying relationships with others.

Dr. Taormina and Dr. Ho also discovered that people who rate higher on scales of extraversion have higher relationship satisfaction. Extraverted individuals enjoy being around others, feel proactive and effective when dealing with issues in a relationship, and actively perform to increase relationship health and satisfaction. Level of agreeableness also positively impacts relationship intimacy and satisfaction. Agreeable individuals are more likely to be supportive, kind, and helpful, and agreeableness can help minimize conflicts and promote positive interaction. Individuals with high scores on the dimension of neuroticism are likely to struggle more with satisfaction within relationships. Individuals high on scales of neuroticism are more likely to worry, doubt themselves, be anxious, and are more sensitive to the effects of negative emotions. Thus, higher rates of neuroticism will make them be more fearful of losing their partner or not being accepted by their partner, which will negatively impact their intimacy and satisfaction in the relationship and increase feelings of jealousy.

Physical attractiveness is positively correlated with relationship satisfaction and intimacy. Not only does attractiveness influence the formation of relationships, but, in quality relationships, it can also help maintain the relationship.

An attractive couple enjoying time together. Physical attractiveness is correlated with higher self-esteem and less anxiety, which lead to better interpersonal communication, greater relationship satisfaction, and less conflict. (Simone Van Den Berg/ Dreamstime)

Physically attractive individuals tend to be more respected, more assertive, more persuasive, and more outgoing, and have higher self-esteem. All of these traits contribute to increasing amount and quality of communication within the relationship and increasing satisfaction and intimacy. High self-esteem specifically has a positive effect on relationships. Feeling good about oneself and having a sense of self-worth and self-acceptance actually contribute to less conflict in a relationship and better communication. Unsurprisingly, shared values also contribute to relationship satisfaction. Having similar values can increase agreeableness and harmony in the relationship. Shared values also contribute to validation of one's beliefs, which is reinforcing.

Gender attributes such as the degree of femininity and masculinity can also contribute to intimacy and satisfaction. Highly feminine individuals are more likely to be caring, attentive to emotions, and expressive of feelings and thoughts. Degree of femininity positively contributes to relationship satisfaction. Masculinity is associated with being less sensitive to others' feelings and less expressive of thoughts and feelings. High masculinity within a relationship can negatively impact intimacy and relationship satisfaction. Healthy relationships need a balance, and relationships with higher levels of sensitivity in both partners tend to have higher intimacy and greater satisfaction.

Interpersonal trust emerges as another important factor for relationship satisfaction. Interpersonal trust means that each individual trusts that the other will be dependable, predictable, and faithful. Having trust in one's partner increases relationship satisfaction. Of course, having trust is only beneficial when the partner is trustworthy. Emotional support and feedback from family outside of the relationship also tends to increase relationship satisfaction. This support from family increases one's confidence, stability, and expectations about relationships and overall contributes to higher intimacy in the romantic relationship.

Having a sense of self-efficacy also contributes to relationship health. For example, having an internal locus of control adds to relationship quality. Locus of control refers to how much control one feels that he or she has over a situation. An individual with an internal locus of control feels as though he or she can control a situation and the outcome. In relationships, individuals with internal loci of control would feel they have the power to improve or change aspects of their relationships and would be more likely to talk to their partners, endeavor to make changes, and actively try to solve problems. Alternatively, an individual with an external locus of control feels like the control in a situation is beyond his or her power. For example, individuals with external loci of control may feel that they have no control over the outcome after an argument occurs. Those with external loci of control are less likely to take responsibility for problems in the relationship, less likely to try to solve problems, and more likely to feel like victims. If both individuals in a relationship have high internal loci of control, the relationship tends to be more stable, more intimate, and more satisfying because both individuals take responsibility and actively attempt to solve problems when they occur.

When examined, engaged individuals tended to have the highest reported rates of relationship satisfaction, followed by individuals who were dating. Married individuals reported the lowest levels of satisfaction on all dimensions of intimacy. Those who are dating or engaged may invest more effort into connecting with their partner and sharing hopes for the future. Married individuals may be more stable and comfortable and expend less effort on their partners. Married individuals also tend to be older and may have other concerns that occupy their attention, such as having children. Having children negatively impacts relationship satisfaction across every dimension. Having children interferes with intimacy, increases stress, and decreases the amount of time spent alone with a partner.

INTIMACY AND SEXUAL SATISFACTION

Attractiveness is important in a romantic relationship but other factors emerge as important for sexual satisfaction. Emotional intimacy, for example, can impact sexual satisfaction for both men and women. Emotional intimacy is independent of physical intimacy, but a strong emotional foundation can positively influence sexual interest, particularly for women. Women who feel accepted, close, and committed to their partners have higher sexual satisfaction. Women's sexuality has become more accepted, especially in Westernized cultures, which adds to sexual satisfaction. Additionally, traditional gender roles are changing. More men are staying home to care for children and previously female tasks such as housework and child care are being more equally divided between the genders, which increases intimacy within relationships. In Norway, a culture on the cutting edge of gender equality, those relationships where housework is more evenly divided were found to have the highest rates of relationship satisfaction and sexual satisfaction from both partners. The shared responsibility and mutual respect led to greater feelings of intimacy and more sexual desire and satisfaction for both partners.

Men's sexual satisfaction is also tied to creativity, variation, and frequency. Men are attracted to women who look happy and with whom they do not have a large number of conflicts. Researchers from the University of Oslo in Norway, the University of Zagreb in Croatia, and the University of Lisbon in Portugal asked men to rate their sexual satisfaction, their emotional intimacy, the sexual attractiveness of their partner, and their satisfaction with the division of domestic duties. Researchers found a direct relationship between fair division of domestic duties and sexual satisfaction for these men. They also found that fair division of housework was correlated with lower conflict, more positive emotions, and the perception of emotional intimacy for both partners, which positively impacted sexual satisfaction.

TO COMMIT OR NOT TO COMMIT?

Being in a committed relationship may affect how individuals view other potential partners but what specifically determines whether an individual stays committed or cheats, and does attractiveness play a role? To examine this question, researchers examined gender, sexual satisfaction, physical attractiveness, relationship length, and social status and found that all of these characteristics play a role in commitment.

With respect to gender, commitment levels vary predictably. Evolutionary pressures have differentially shaped the sexual behavior of men and women. As previously noted, women bear the risk of reproduction to a greater extent than males. As a result of this increased risk, women tend to endorse higher levels of commitment and are less likely to engage in sexual intercourse outside of a committed relationship. This means that men and women approach dating and sexual relationships differently. Men report a desire to have upward of 20 sexual partners over their lifetime. Women, alternatively, report the desire for fewer than five. Men prefer sexual variety and are more likely to consent to sex more quickly, and women prefer sexual quality and will wait to ensure commitment and value. Men were more likely than women to agree to go to a stranger's apartment or to agree to have sex when approached by a stranger in a psychological study. Men are also more likely to agree to sex without emotional intimacy, are more likely to engage in extramarital affairs, and felt more comfortable after a one-night stand than women.

Obviously not all men cheat (and not all women are faithful). There are many variables that contribute to individual behavior, including life history, personal values, and relationship goals. When considering an extramarital relationship, projected length of the alternate relationship impacts preferences and behavior. New long-term relationships carry costs for both partners, and, thus, standards are high. When considering a new long-term relationship, both men and women would need to carefully assess the costs and benefits before making a decision. As discussed in Chapter 5, short-term relationships carry fewer costs for men, and thus they tend to be more open to engaging in sexual behavior. Men are more likely to choose to engage in a short-term relationship while maintaining their current relationship. Short-term relationships for women, however, carry high costs, maybe even higher costs, because she can become pregnant, and if she risks the trust of a long-term partner, she would bear the costs all alone. Thus, it could be predicted that women would have higher standards for short-term relationships and would be less likely to cheat unless the new partner was of exceedingly high status or attractiveness.

Tobias Greitemeyer from Ludwig-Maximilians University in Germany examined the differences in the decision-making process for men and women in a situation where they encounter a new potential partner. He proposed that women and men might place different values on different characteristics. For example, in line with the attractiveness research, Dr. Greitemeyer predicted that men would likely value physical qualities more than women and women may value wealth and status more than men in a potential new partner.

This prediction is not surprising, given that we have already discussed that research overwhelmingly shows, cross-culturally, women are attracted to partners who can provide resources and protection and men are more attracted to physical cues of fertility. In Dr. Greitemeyer's study he not only expected that men would be more likely to value physical attractiveness in a new potential partner and women would be more likely to base their decisions on the reported socioeconomic status of the potential mate, but also expected that men would be more likely to respond to a sexual offer than women. Additionally, he expected that men would only consider physical attractiveness in their decision, while the criteria women used to make their decision would vary based on the type of relationship. Specifically, for short-term affairs, women would include physical attractiveness in their assessment, but for a long-term commitment, women would make their decision based on socioeconomic status and attractiveness would be less important in the woman's decision.

In a hypothetical scenario, Dr. Greitemeyer asked participants to imagine being in a relationship with another person. They were provided with a photograph and information about that partner. They were next presented with a highly attractive alternative and an alternative with a reported high socioeconomic status. They were then asked whether they would agree to date, kiss, make out with, have sex with, or leave their current partner for the potential new partner. Males were more likely to endorse engaging in each of the described behaviors when the potential partner was physically attractive and less interested when the partner was of high socioeconomic status. Females were more likely to accept the offers when the potential partner had a high socioeconomic status. Furthermore, men were more likely to endorse kissing, making out, and having sex with a new partner than were women. Men were more swayed by physical attractiveness and women gravitated toward higher economic status, particularly if considering being in a relationship with the potential partner. For kissing, making out, and short-term sexual intercourse, women also included level of attractiveness in their decisions (in addition to socioeconomic status), making women's requirements even higher (high socioeconomic status *and* high attractiveness). Thus, both physical attractiveness and socioeconomic status play a role for women. Men, alternatively, preferred an attractive potential partner but were actually repelled by females with high socioeconomic status.

When encountering a potential new partner, one must decide whether to stay faithful to the current partner, cheat, or end the current relationship. When meeting a potential new partner, it makes sense that one may evaluate the possible benefits and differences between the potential partner and the current partner. If this evaluation leads us to deduce that our current partner is more attractive, has higher social status, or is otherwise better than the potential partner, then it is easy to remain faithful. If, alternatively, the new potential partner offers qualities that our current partner does not, such as higher attractiveness, higher status, less conflict, or other positive qualities, then we may believe the potential partner is better and act accordingly.

ATTRACTIVENESS AND DIVORCE

Although continued commitment in a monogamous long-term relationship may be a Westernized ideal, sustained relationships are not a certainty. Many relationships, even long-term relationships, may eventually end regardless of the initial intentions of the individuals involved. In the United States, about 30 percent of marriages end within the first 10 years and only about 55 percent of marriages last 20 years. This means that almost half of marriages in the United States eventually end. There are many factors that influence the likely length of a marriage including age at time of marriage, education level, income, location, population, attractiveness, and experience, among other factors. Research shows that those who marry before the age of 25 are more likely to get divorced. Those who marry later in life tend to have more stable relationships and stay married longer. Those with a college degree are also about 10 percent less likely to get divorced and higher incomes are associated with less relationship stress and more stable relationships overall.

Couples with children are more likely to stay together, but they are not necessarily happier or more satisfied with the relationship. Relationship or marriage satisfaction actually tends to sharply decline after the birth of a child and only slowly recovers as the child matures and moves away. Couples tend to experience higher rates of stress (and less sleep), more financial struggles, and more negative exchanges when caring for young children. Couples report being less sexually attracted to one another and more irritated with one another after having children. Couples who decide to have children to solidify and strengthen their relationships are misguided. Children have many positive qualities, but improving marital satisfaction is not among them. Strong couples who have good communication, financial stability, and good cooperative skills tend to remain the most stable after starting a family.

Factors such as shared religious faith, family circumstances, and location also influence divorce rates. Shared religious faith tends to lead to fewer divorces. Having shared faith and values can increase cohesion within a relationship. Many religions also discourage divorce, so highly religious couples may stay together regardless of relationship quality. Family circumstances, such as having divorced parents, also influences divorce rates. Children of divorced parents are more likely to get divorced themselves. Furthermore, couples who live together before marriage are also likely to divorce. Although this finding may seem counterintuitive, it is likely that these less traditional individuals are more likely to end a relationship if it is no longer mutually beneficial, whereas more traditional couples will stay committed and attempt to work through problems. In a society where the cultural ideal is lifelong monogamy, such findings are disheartening. However, in many circumstances, these relationships are no longer satisfying or mutually engaging, and divorce may allow for the development of more rewarding future relationships.

Divorce rates vary substantially across cultures. Divorce rates are high in the United States, but they are even higher in Spain, Portugal, Hungary, and many other countries around Europe where 60 percent or more of marriages end in divorce. Divorce rates peak in Belgium with 70 percent of couples eventually splitting up after marriage. In other areas of the world, divorce is very uncommon. In Mexico, for example, only about 10 percent of married couples eventually divorce. In Chile, only about 3 percent of marriages end in divorce.

What is the driving force behind divorce? Research shows that the reasoning behind divorce varies by gender. Women are more likely to file for a divorce from males who fail to provide. Men are more likely to file for a divorce from females who are sexually unfaithful. Since one of the most attractive features about a man is his willingness to provide resources, if he does not do so or if he becomes lazy and complacent, women may be driven to move on and secure a more stable partner. Since one of the most attractive features about a woman is her ability to produce children (whether that is the goal of the relationship or not), sleeping with another man (and thus compromising the genes of the next generation) is the biggest deal breaker for men.

David Buss, from the University of Austin, illustrated these gender differences within relationships by examining the causes for breaking up with a partner. Women were found to be much more likely to forgive a partner's

An older couple nap together while awaiting a flight at the airport in Santiago de Chile. In Chile, only 3 percent of marriages end in divorce unlike countries such as Belgium where 70 percent of marriages eventually end. (Dan Fleites/Dreamstime)

sexual infidelity but not likely to forgive emotional or financial infidelity. Men, on the other hand, were less concerned about their partners having emotional ties with others, as long as they did not engage in sexual behaviors with other men. In this study, participants were asked to either imagine their partner engaging in a deep emotional attachment to someone else or enjoying passionate sexual intercourse with another person. While neither gender enjoyed thinking of these circumstances, women were more upset by their partners creating an emotional bond with another woman and men were more upset by their partners engaging in sex with someone else.

Dr. Buss found that sexual infidelity, while provoking for women, was actually rated as more forgivable, particularly if the male was repentant and had no emotional bond with the other woman. Having an emotional bond meant risk of the partner bestowing physical resources on another person and potentially ending the current relationship. One-time sexual infidelity devoid of emotional intimacy was less threatening. For men, the opposite was true. A man was more likely to forgive his partner for having an emotionally intimate connection with another man, as long as she was not engaging in physical intimacy. If the female partner has sexual relations with another man, even if the actions are devoid of an emotional connection, men are much more upset. Even one-time sexual relations increase the risk of her bearing another man's child, a potentially relationship-ending offense.

Marriage has been found to have positive effects such as increased life expectancy, better health, higher incomes, more support, and higher life satisfaction. Marriage is also associated with weight gain and decreased objective attractiveness. However, Dr. Petter Lundborg and colleagues from Lund University in Sweden found a connection between divorce rates and body mass index (BMI). In general, while marriage is associated with weight gain, divorce is associated with weight loss. Thus, Dr. Lundborg and colleagues proposed a correlation between BMI of married adults and divorce rates. They proposed that the national divorce rate would be correlated with BMI of married people. They found that in countries where divorce rates are high, married people were more likely to take care of their bodies and maintain lower BMIs and stayed fit and competitive for the dating scene. Thus, marriage may be correlated with a decrease in efforts to maintain physical attractiveness, and risk of divorce creates the need to remain physically attractive to secure a new partner. Thus, there is a correlation between average BMI and divorce rates across cultures. Cultures with higher divorce rates have lower average BMI. Since this is not a causal study, it is unclear whether maintaining physical appearance leads to great incidence of divorce or if the risk of divorce influences people to maintain their physical appearance. Certainly, either could be the case. Maintaining physical appeal would lead to higher self-esteem, increased mate-poaching attempts from others, and more potential jealousy from a partner. However, concern about divorce could also provide motivation to maintain physical appeal.

FLIRTATION

When men or women find another individual to be physically attractive, they are likely to engage in flirtatious behavior to communicate interest. Men and women tend to engage in different behaviors when flirting, but these behaviors are quite consistent around the world. When flirting, women engage in more smiling behavior, and tend to lift their eyebrows, open their eyes more widely, and make more eye contact. After engaging the partner, women then tend to look down, tilt their heads to the side, and then glance away. Women also tend to cover their faces with their hands and laugh coyishly. This behavior communicates interest and femininity and invites the masculine partner to approach and engage.

Men, alternatively, all over the world, tend to accentuate masculine behaviors when interested in a woman. They tend to take up more physical space with their bodies, thrust their chests out, stand taller, and behave in a more extroverted fashion. Interestingly, these same behaviors are used when trying to intimidate and express dominance over other males as well as to demonstrate interest and availability to a woman, likely complementary goals.

Flirtation can serve to announce availability or to express interest in another person. If people are already in a relationship, being the recipient of flirtatious advances can cause them to reconsider which alternative they prefer. In this way, flirtation from others can decrease current relationship satisfaction because it causes assessment of the positive and negative aspects of the current relationship. For example, when an attractive woman flirts with a man, the man is more likely to report being less satisfied with his current relationship. Men and women who are the object of flirtatious behavior are more likely to find fault with their partners and less likely to forgive their partners for small offenses. Being aware of attractive alternatives can raise the standards for the current partner. Women, however, are more likely to attempt to ignore the flirtatious behavior and direct their attention to working on strengthening their current relationships rather than formulating new ones. Research has demonstrated that men are more likely to respond to flirtatious behavior but that men who want to maintain a committed relationship can be trained to ignore flirtatious behavior and to view flirtation as a threat to their current relationship. Women do not seem to need such training.

SEXUAL FANTASIES

Just as flirtatious behavior differs between men and women and can impact relationship satisfaction, the differences in the process of attraction between males and females are apparent in sexual fantasies as well. Researchers examined the content of sexual fantasies between males and females and found consistent differences in the content of such fantasies that varied predictably by gender and by level of attractiveness.

For example, the dichotomy illustrating the different mating strategies of men and women shows itself in the content of sexual fantasies. When empirically studied, men have twice as many sexual fantasies as women. The content of male fantasies is more likely to revolve around having multiple sexual partners, varied partners, anonymous partners, group sex, and sex with attractive strangers. The focus of male sexual fantasies tends to revolve around feelings of lust, physical gratification, and physical attractiveness of the partner. Due to differences in the level of parental investment, it makes evolutionary sense for men to copulate with more partners as a possible means of increasing their overall reproductive success. In sexual fantasies, their focus is more on the physical acts of sexual intercourse than the emotional intimacy that sex can facilitate.

In contrast, women tend to have fewer sexual fantasies as compared to men. Women usually fantasize about someone they know rather than about varied, anonymous strangers. The focus of female fantasies tends to be on personal and emotional tenderness and aspects of romance rather than physical gratification. Women focus on the relationship, the emotional intimacy, and the sharing of the experience. Female fantasies focus more on the emotional connection created rather than on the physical aspects of sex.

Additionally, men are more likely to masturbate than women and tend to fantasize more during masturbation. Over half of men report that they think about sex and masturbate daily. Fewer than one-fifth of women think about sex and masturbate so frequently. Both genders have internally triggered fantasies, but men's fantasies are more likely to be triggered and revolve around things that they see or hear in the external environment while women's fantasies tend to be triggered by thought, feelings, memories, or other internal stimuli. It may be the case, however, that men have more externally triggered fantasies simply due to self-exposure to explicit materials. Men are more likely to seek out or orient to pornography or sexually explicit material than women.

Exposure to sexually explicit materials or engaging in sexual behaviors influences the production of testosterone in men and women. Sexual thoughts and sexual activity with a partner increase testosterone production in women while visual erotic stimuli increases testosterone in men. Sexual activity can include physical attraction and physical contact or emotional connection. Explicit physical contact is linked to increases in testosterone while nurturing or emotional interactions decrease testosterone. Nurturing aspects can include aspects of relationships like romance, support, affection,

The Impact of Attractiveness on Behavior and Relationship Satisfaction   135

and warmth. Many of these aspects are present in sexual fantasies, so some fantasies may actually lower testosterone levels. When asked to read and think about having an explicit sexual encounter, testosterone levels for women (but not men) measurably increased. It may be the case that small increases in testosterone have a greater impact in women because they have naturally lower levels. For men in this situation, the increase in testosterone may not be distinguishable from their already high levels.

Men are more likely to describe sex in terms of physical acts and pleasure. Women are more likely to describe it as connected with love, commitment, and relationship satisfaction and maintenance. For both men and women, common fantasy themes are imagining past experiences, imagining sex with one's current partner, or imagining sex with an attractive other. Women's fantasies tend to have more emotional content, are built around romance, and focus on personal characteristics of their partner as compared to the fantasies of men. Women also tend to fantasize about commitment and do not tend to place themselves in the role of the aggressor or instigator of the sexual behavior. They do, however, emphasize their own pleasure. Men tend to imagine more explicit sexual contact and content than women. They include more descriptions of sexual organs and revolve around the actual intercourse rather than the foreplay, romance, and aftermath.

PORNOGRAPHY USE AND ITS EFFECTS ON SEXUAL AND RELATIONSHIPS SATISFACTION

Pornography is content designed to elicit sexual arousal and interest. Pornography can supersede the constraints of reality and provide the image of an ideal sexual experience that is not bound by realistic relationships and consequences. Pornography goes beyond individual sexual fantasies and is essentially a collective fantasy. The intent of pornography is to elicit sexual arousal. Thus, there must be a goodness of fit between the viewer and the content. Due to differences in what each gender finds to be the most attractive, men and women are aroused by different stimuli and the exposure has different effects on relationships and sexual satisfaction.

In the United States, the pornography industry made an estimated $13 billion in 2015, out of which $3 billion alone came from the Internet. Although men are 543 percent more likely to look at pornographic content than women, more and more women are watching pornography with 18 percent of women admitting to viewing pornographic content when asked during a psychological study. There are varying extremes of attitude toward pornography. Some find it sexually appealing while others claim that it is corruptive, derogatory, and violence inducing, and increases the likelihood of rape. No empirical evidence has suggested support for these negative effects. Studies did find that pornography use increases reports of masturbation and sexual exploration and tends to have a negative effect on overall relationship and sexual satisfaction.

There are empirically supported differences in pornography consumption between males and females. The bulk of the consumers of the pornography industry in the United States are males. Only between 10 and 20 percent of women seek out pornography on a regular basis. Men tend to find pornography more sexually stimulating and exciting and are more drawn to hard-core, sexually explicit content. Men prefer that the women in the pornographic content are young and physically attractive. Women, on the other hand, tend to relate to soft-core content that focuses more on emotional and psychological arousal. Women tend toward porn that looks genuine, has a story line, and has slow buildup with soft images, lighting, and music. Women are less concerned about the physical attractiveness of the actors and more concerned about the emotional connection. Males tend to be more interested in the physical acts of sex, but women are in it for the buildup.

There is some research that suggests that viewing pornography increases sexual empowerment and sexual independence. Watching pornography frees the viewer from concerns about their own attractiveness and gives them confidence to explore and experiment with new sexual acts, either by themselves or with a partner. It also allows individuals to view sex as less shameful. Dr. Martin S. Weinberg and colleagues from Indiana University studied college students of both genders to examine their use and interest in pornography. They analyzed 101 women and 71 men of hetero- and homosexual orientations. Some positive outcomes of pornography use were that all of the groups reported that watching pornography gave them a more expansive understanding of different sexual acts and increased their interest in engaging in new sexual experiences (such as watching others or being watched, having sex with multiple partners or in multiple ways). The groups reported being more interested in sex toys, anal sex, and oral sex. They also tended to engage in more sex with more partners, and engaged in more self-stimulation behaviors after watching pornography.

Use of pornography, however, was also found to have some negative consequences for psychological health and relationship satisfaction. Watching pornography provides unrealistic expectations about intimate relationships and is negatively correlated with sexual and relationship satisfaction. An interview study found that the more college men viewed pornographic material, the more concerns they had about sex. Men who view pornography frequently report less enjoyment when engaging in sex with a partner and have altered expectations about sex and behavior. Men who view pornography are likely to request that their partner recreate pornographic sex acts, need more pornographic imagery to maintain arousal, and worry about their own performance and body image. Use of pornography within a relationship is correlated with lower ratings of intimacy and relationship quality. Females in these relationships tend to have feelings of inadequacy and experience lower self-esteem than women who are not in relationships that use pornography. Pornography use also tends to decrease men's judgments of their partner's physical attractiveness, causing their partners to report greater concerns and more negative feelings about body image.

Many young people learn about sex through pornographic content on the Internet. Most boys and girls will be exposed to pornography, either accidentally or by choice, by age 16, prior to the average age of sexual experience. Thus, entire generations may develop with pornography as their source for sex education, shaping their expectations about sex. This education may portray unrealistic or uncommon sexual practices and also tends to omit information about health consequences, safety, or possible pregnancy. Use of pornography, particularly during adolescence, may also contribute to adolescents' cognitive scripts of what sex should be, how they should behave during sexual encounters, and how their partner should look and respond. In addition to omitting the possible negative consequences of sex, pornography also excludes the importance of building intimacy between two people. Although parents of these adolescents report they do not support their teen's pornography use, they are underprepared to discuss it with them.

Dr. Jane Brown from the University of North Carolina and colleague Dr. Kelly L. L'Engle from Family Health International further examined the effects of pornography exposure on adolescents' sexual development. Some adolescents were at higher risk of exposure to pornographic material. These included black adolescents who were sensation seeking, were from households with parents who were less educated about sex, and were from a lower socioeconomic status. They found that these adolescents who were exposed to pornographic material early in their adolescence had less progressive gender role attitudes as they entered adulthood and were less likely to use contraceptives. Adolescent males who viewed pornography in magazines, in movies, or on the Internet were more likely to have broader sexual norms and were more likely to engage in sexual harassment later in life. Both males and females who were exposed to pornographic material early in adolescence were more likely to engage in oral sex and sexual intercourse at earlier ages. Education via pornography also has the potential to be damaging because almost 90 percent of scenes from the most widespread pornographic videos contain physical aggression. These acts are overwhelming perpetuated by men toward women.

Dr. Chyng Sun from the School of Professional Studies in New York and colleagues from Arkansas and Virginia examined specific effects of early pornography use on sexual behavior in male college students. They found that a majority of the college-aged men in their study used pornography on a weekly basis, mostly through the Internet, and that almost half were exposed before the age of 13. Their study revealed many of the less favorable effects of viewing pornography. Upon analysis, use of pornography for this group of college men was associated with increased reliance of pornographic material to achieve and maintain arousal, more pornographic acts during sex, and a weaker intimate connection with a partner. Although the exposure to pornography did not negatively affect their sexual security or body image, it did correlate with less overall sexual satisfaction and poor relationship satisfaction. Furthermore, exposure to pornography correlated with more sexual experience, earlier age of first sexual experience, and more mimicking of pornographic acts during sex.

Thus, watching pornography can stimulate positive feelings of arousal but it also has the potential to impact body image of the viewer, make one more judgmental of their partner's level of attractiveness, and decrease the level of intimacy and satisfaction within a relationship. Couples who view erotic or pornographic materials with one another were found to have better communication about sex and fewer sexual problems and reported a tendency to engage in sexual activity following the viewing of the materials. Repeated viewing of sexually explicit material, however, particularly without one's partner present, had negative effects on the ratings of their partner's level of attractiveness, sexual appeal, performance, and curiosity. If the couples in the explicit material are attractive, viewers may also feel more negative about their own appearance or the appearance of their partners.

Almost three-fourths of adults report interest in pornography and more women are reporting an interest over time. Additionally, an individual's interest in pornography tends to change after committing to a relationship. After marriage, men report viewing less pornography and women report viewing more. This is likely because, during marriage, most pornography is viewed together rather than alone. Those who view pornography report more interest in unconventional sex practices than those who do not.

Ideal pornography content for males depends on attractiveness and physical behavior of the female actor. The attractiveness of the male actor did not factor in to female arousal. Men are exposed to pornography at younger ages and tend to view it more frequently, and more often by themselves. Men prefer sexually explicit pornography more frequently than women, and women are more likely to watch pornography if they are watching with a partner. In a study completed by Dr. Gert Martin Hald at the University of Copenhagen, significant gender differences emerged on the preferred content. Although both genders endorsed watching pornography that represented vaginal sex, men preferred to watch anal sex, oral sex, group sex, lesbian women, and amateur sex more than did women. Women preferred soft-core pornography focused on emotional and psychological arousal more than men did.

Overall, pornography use is correlated with earlier and more frequent sexual engagement, higher rates of multiple sex partners, poorer ratings of intimacy between partners, lower sexual satisfaction, and more frequent anal sex and use of drugs or alcohol during sex.

CONCLUSION

One's level of attractiveness has widespread implications for many aspects of relationships and behavior. Level of attractiveness is correlated with relationship length, quality, intimacy, and satisfaction. Attractiveness also influences one's level of commitment to a partner and high levels of attractiveness are correlated with more jealousy, flirtatious behavior, and rates of divorce. Attractiveness factors in to male and female sexual fantasies, impacts sexual confidence, impacts sexual satisfaction, and is inherent in pornographic content. Attractiveness elicits expectations and behavior from others with respect to the length and type of relationship sought. It also provides information about likely future behavior and likelihood of commitment or cheating behavior.

Although rating high on scales of attractiveness has widespread social and biological benefits, it can also indicate less of a likelihood of maintaining a long-term faithful relationship, particularly for males. High levels of attractiveness increase mate-poaching attempts and increase interest and opportunity for extramarital relationships. This decrease in commitment can impact mating opportunities, intimacy, sexual satisfaction, and relationship satisfaction.

7
Psychological Effects of the Preoccupation with Beauty

Given the vast array of research on the topic of attractiveness and the multitude of positive and negative outcomes that result from one's level of attractiveness, it is not surprising that many people are preoccupied with their own and other's physical appearances. People in the United States, in particular, live in an extremely body-conscious society, and one's perceived level of attractiveness can have substantial ramifications on one's confidence, appeal, motivation, opportunities, and expectations. These ramifications can influence personal relationships or academic or professional achievements.

Based on our culture's extensive preoccupation with physical appearance, research has been conducted on the effects of being body conscious. Chapter 2 outlined the benefits of being attractive but there can also be negative ramifications that results from the concern for one's own appearance, particularly if it does not match one's desired appearance. These ramifications include effects on body image, self-esteem, behavior, and psychological and physical health. Because media plays such an impactful role on shaping beauty standards, this chapter will first examine the role of media on self-perception and then will look at research on the possible impacts on body image and psychological and physical health.

Common results of exposure to media that normalizes extreme examples of beauty include impacts on body image, self-esteem, and self-worth, and development of eating disorders, depression, and anxiety, for men and women. Over the span of the first 10 years of the 21st century, rates of depression for females in the United States doubled, and almost all teenaged girls studied expressed displeasure with their own appearance. Between the ages of six to eight, 42 percent of girls studied endorsed wanting to be thinner. By age ten, 51 percent of girls report dieting to lose weight. By age thirteen, 53 percent admit to being unhappy with their appearance, and this rate increases to 78 percent by the end of the adolescent years. In adulthood, 90 percent of women rated that they see themselves as too fat or too thin and report that they have taken measures to change their bodies. Research reveals, however, that not all women experience these negative effects in the same way. Race, class, and sexuality play a role in how women view themselves with respect to the cultural ideal. For men, media exposure has similar effects. Exposure to unrealistic ideals is correlated with poor body image and lack of sexual interest, as well as the development of depression, anxiety, and eating disorders.

INFLUENCE OF THE MEDIA ON BEAUTY AND BEHAVIOR

There tends to be universal and cultural expectations about attractiveness. Since cultural expectations are not innate, one must learn what is most attractive in a particular society. What we should look like and what we should achieve is communicated to us through family, friends, society, culture, and media from birth and throughout our lives. Such learning occurs through positive reinforcement in personal interactions as well as through media advertisements. Some rewards of attractiveness are naturally occurring. Catching the eye of the opposite sex, garnering the attention in groups of people of both sexes, and being rewarded and given opportunities by others are naturally reinforcing. In addition to these natural consequences of attractiveness, cultural advertisements further inundate us with what we should look like, what we should find attractive, and how we can achieve such ideals.

Research demonstrates that there are ubiquitous social pressures on individuals to conform to attractiveness standards of the culture in which they live. Those who do not conform will likely encounter at least gentle pushback from other members of the society and have been described as lazy, as outsiders, and as undesirable. Society also tells us what we should be attracted to, and, while individuals can ignore or resist cultural norms, many will encounter social scrutiny, confusion from others, and both subtle and overt pressure to conform. Some individuals may be strong enough to withstand this pressure, but others quickly align their behavior to fit in.

For example, in the United States, friends and strangers alike will react when noticing a woman who chooses to not shave her armpits. Shaving behavior is a personal choice but others will notice, judge, and distain, and may display shock or even disgust when the norm is violated. Some successfully withstand or ignore this extra social attention, but more frequently, individuals simply consciously or unconsciously conform to avoid ostracism and scrutiny. Thus, individuals tend to conform to many beauty standards due to social pressures. Some of the typical behaviors and practices contribute to maintaining health, some advertise or mimic genetic quality, and some are merely a by-product of aligning with societal norms, which are arbitrary and shift over time.

Although different cultures have historically had their own beauty ideals and practices, Westernized ideals about beauty have been seeping into other markets in more recent generations. In the Internet age, information can be shared instantaneously, and other cultures have more exposure to pervasive media from the United States, including advertisements and ideals. With this exposure, other cultures have shifted their practices as well as the content of their media. A study conducted by Seung Yeob Yu from Namseoul University revealed that almost 20 percent of models in Korean magazines are Caucasian women, and, in Chinese fashion magazines, 75 percent of the cosmetics advertised are not Chinese brands. Thus, the cosmetics and beauty standards for Asian women have shifted to display ideals from international and American companies.

Almost a third of the models in Chinese magazines are Caucasian women, demonstrating that white models are considered highly beautiful, even in an Asian culture. The advertisements also emphasize self-confidence, distinction, and liveliness rather than the traditional Chinese values of goodness, poise, and grace. So, if chasing beauty ideals is difficult for the average American woman, imagine how difficult it is to conform to a standard for a different ethnic group. Attempting to conform to a Caucasian look requires skin bleaching, hair transformations, and facial alterations beyond cosmetics. This has led Asian men and women, in particular, to seek out cosmetic surgery in numbers greater than ever before.

Thus, media, particularly media from Europe and the United States, puts forth a standard of attractiveness that is ubiquitous throughout the industrialized world. Media supposes an ideal that informs individuals about whom they are supposed to be, particularly with regard to physical appeal. Media capitalizes on attractive individuals in an attempt to sell products and catch the attention of the viewer. As a side effect, media undoubtedly plays a significant role in shaping individuals' perceptions of their own bodies. The average teenager spends at least 80 hours a week watching TV, listening to music, reading magazines, surfing the web, and playing video games. Such pervasive exposure to the media shapes their perceptions of ideals during a period where their bodies are changing and they are finding their place in the adult world.

A Chinese woman walks up a flight of steps in Shanghai plastered with a fashion magazine advertisement portraying a Caucasian model. Such exposure and internalization of Caucasian beauty ideals has led to increases in skin whitening products and cosmetic surgeries throughout Asia. (AFP/Getty Images)

Fashion Magazines

Media tends to highlight the most ideal forms when selecting models. Fashion magazines, television ads, video game characters, social networking sites, and even animated images tend to sexualize the female form and focus on appearances rather than internal qualities or productive abilities, leading to objectification of the female body. Research demonstrates that simply looking at these magazines or viewing such programs serves to increase the extent to which these cultural ideals are internalized. Repeated exposure to such material changes the zeitgeist of a culture. Exposure teaches women and men what to value in themselves and what to look for in others. Since models for such fashion magazines, in particular, portray very select and extreme characteristics, the average person may struggle to meet the ideal. This difficulty in meeting the ideal can affect self-confidence and body image and can also be damaging to physical and mental health. Exposure to such images has been shown to increase self-objectification, causing both genders to be hyperfocused on their own appearances and preoccupied with how they may be viewed by others.

Examination of current media trends shows that there is a particular focus on the sexuality and objectification of women and girls. Overall, the trends for female beauty in print magazines include increasing thinness, displaying the entire body in photos, and increasing skin exposure. Although in psychological research, waist-to-hip ratio emerges as more important for the ratings of attractiveness of females by males, fashion magazines promote exceptionally low body mass as a characteristic feature for their beauty models. Mia Foley Sypeck from the American University in Washington, D.C., analyzed feminine beauty as portrayed in fashion magazines from 1959 to 1999. Over this time frame, models have become exceedingly thin and more exposed, and now full-body photographs are more common than they were 50 years ago. From the 1960s on, the ideal weight for female models has been becoming progressively lower. In the 1980s, the ideal female model shifted from being just thin to being muscular and toned. This shift from just thin to thin and toned may present a healthier body type, but it makes the ideal even harder to reach. Now, to meet the cultural ideal, women must not only be thin, but also toned, muscular, and fit.

Currently, the ideal female body weight portrayed in fashion magazines is 23 percent below the average for a typical woman's weight. Such a discrepancy has widespread negative impacts on how women view themselves and creates a mismatch between a woman's actual body and her expectations for her ideal body. It also shapes how women view others and how men view the typical female form. Furthermore, these underweight models still typically have a waist-to-hip ratio within or under the ideal, or else their images are altered until they do. Increased exposure to these models by average women provides women with an unrealistic ideal of what the female body is supposed to look like. This enhanced ideal of thinness coupled with the modern stigmatization of obesity is likely a predominant factor in the radical increase in eating disorders in the modern era, as discussed later in this chapter.

In this sense, fashion magazines not only influence how individuals view others, but also influence how individuals view and feel about themselves. These self-evaluations of attractiveness are one of the strongest predictors of self-esteem, particularly for young American women. Being considered physically attractive positively influences feelings of confidence, self-presentation, self-concept, and ability to create a strong first-impression, and increases the likelihood of being selected for friendship or as a romantic partner. Since the cultural ideal of female beauty pervasive in media is an extreme subset of women, average women may struggle to maintain the high confidence and positive self-esteem needed to be successful in the social and professional world.

The effects of exposure to fashion models have been long debated, and research has revealed that exposure to such media is linked with negative psychological effects and behaviors. Women tend to be less comfortable with their own bodies when self-comparing to media ideals. Leora Pinhas and colleagues from the University of Toronto empirically examined the effects of such media on mood and body satisfaction. In their study, women were exposed to pictures of models in fashion magazines. After viewing photos of these models who represented the thin ideal, women were more depressed and angrier than control groups. The images had an immediate effect on mood and took effect after only viewing 20 images.

Television and Movie Actors

Other forms of media have similar impacts on mood and psychological health of the viewer. For example, Leslie J. Heinberg from Johns Hopkins University School of Medicine and J. Kevin Thompson from the University of South Florida both examined the impact of TV commercials on female body image. Women who were sensitive to body image were rated as significantly more depressed and more dissatisfied with their bodies following exposure to thin and attractive models in television commercials. Individuals who have high sociocultural awareness were rated as angrier following exposure to thin and attractive models. So the characteristics of the individual viewing the media matters and media affect different people in different ways. Those with preexisting sensitivity or psychological disturbances may be at the highest risk of responding negatively to the stimuli, and the length of viewing time is linked with the likelihood of internalizing the ideals.

Laura Vandenbosch and Steven Eggermont from the Leuven School for Mass Communication Research in Belgium sought to examine how such exposure impacts women's self-conceptions. They found that exposure to media increased subjects' self-awareness of their own bodies and led them to be more self-critical and to engage in more comparison with others. Such exposure also led to increased levels of anxiety and body dissatisfaction. In their study, exposure to media ideals also contributed to rates of psychological disorders such as eating disorders. The effects were especially detrimental for those individuals who internalized and identified with the standards and for those in the transitional period of adolescence.

Much of the research on the potential negative effects of media has been concentrated on females and female body image. However, there are positive and negative effects on male viewers as well. Laramie D. Taylor and Jhunehl Fortaleza examined the effects of exposure to a muscular ideal for college-aged men. Exposure to a muscular ideal increased body anxiety and increased reported desire to engage in body modification. Exposure to television and movie actors who engaged in violent behaviors also increased the male viewer's body anxiety and decreased his self-ratings of attractiveness. Since actors and images in movies and video games tend to include males with high muscularity and amplified aggressive behavior, their impact on men is worthy of study. Dr. Taylor and Dr. Fortaleza found that exposure to media violence caused decreases in body perception, attitudes, and mood for men. Even reading about violent characters was enough to cause decreased self-ratings of attractiveness, increased anxieties about appearance, and increased desire to develop more muscle mass.

When men and women were asked to rate images for the most attractive body types, they identified those images that are typical of those portrayed in the media. Ideals for males include medium, average frames, muscular large frames, and large broad shouldered frames. Ideal for females included medium-sized, yet lean frames, and small, average builds. Women tended to be more focused and attracted to lean body types for both genders, likely due to being hypersensitive to the pervasive thin ideal for women.

When primed to be self-conscious about one's body, psychological disorders are more likely to occur. For example, research has revealed a direct effect of self-objectification on the development of depression, particularly for women. Objectification is the idea that women are only important for their bodies. Women who are objectified may internalize this idea and become preoccupied with their own appearances, making them more self-conscious. The more frequently a woman is objectified or witnesses objectification, and the more she is exposed to media ideals, the more likely it is that she will self-objectify. This increased self-consciousness, if linked with an inability to meet the cultural ideal, may lead to shame, feelings of inadequacy, or anxiety.

This anxiety may be due to not meeting social expectations or due to feelings of not being safe from onlookers. Women, in general, tend to engage in much more self-monitoring and appearance checking than males and are more likely to become preoccupied with their physical appearance at the detriment of accomplishing and focusing on other tasks. This preoccupation can lead to psychological disorders such as eating disorders, anxiety disorders, body shame, low self-esteem, and depression, particularly among women and during adolescent development.

Video Games and Music Videos

Many different types of media influence behavior and attitudes. For example, violent video games have been correlated with higher rates of delinquency and aggressive behaviors, in both men and women. Due to this documented connection between video games and behavior, Elizabeth Ruth Wack from the University of Florida was interested in discerning whether video games had a similar connection with perceptions of attractiveness. The visual displays and graphics of video games contain unrealistic body ideals for male and female characters. Males tend to be hypermuscular and females are endowed with large breasts and skimpy clothing, and are commonly borderline pornographic. Early studies demonstrated that men who watched a television show with highly attractive female characters rated average women as less attractive than did men who had not watched the show. Dr. Wack proposed that exposure to the graphics in video games may have the same effect. Exposure to such media is likely tarnishing males' expectations and ratings of real women.

Dr. Wack was interested to find whether video games have the same effect as television portrayals on the ratings of female attractiveness. Furthermore, since the male characters in the game are frequently hypermasculinized, she was also interested in the effect the game play had on men's own body images. Previous research has demonstrated that viewing television programs with ideal male physiques decreased males' levels of satisfaction with their own muscles and increased their levels of depression. In her study, Dr. Wack asked males to rate photos of average women for attractiveness and to self-report on their own body satisfaction following exposure to a video game. Interestingly, Dr. Wack did not find negative influences on the players' perceptions of female attractiveness following exposure to the hypersexualized images. Playing video games did not seem to have the same effect as TV on skewing a man's perception and rating of normal women. However, there was a negative ramification from game play on the male player's own self-esteem. Males who compared themselves to the characters in the game were more likely to be dissatisfied with their bodies, more anxious about their bodies, and more likely to endorse measures to change their physical appearance.

Female beauty is used to sell consumer goods to both women and men. The display of female beauty informs people around the world about what is beautiful. When examining music videos, the tendency to portray women as sexualized and available emerges as a predominant theme. Analysis of music videos shows that women are frequently scantily clad and the focus of the video is on their bodies rather than on their faces. Researchers who analyze the content music videos across many musical genres present data that the way women are portrayed in music videos contributes to sexual objectification. Videos portray women as highly sexualized and as objects to satisfy and entertain men. Music videos serve as an exemplar as how media sends messages to girls about who they should be, what they should look like, and how they should act to attract the attention of a man. Similar to other types of media, exposure to music videos, particularly during adolescence, can impact mood, behavior, and self-esteem for women and girls.

The Impact of Photoshop

Research has demonstrated that self-comparison with ideal images, whether in print media, television shows or campaigns, or music videos, can lead to negative mood, body-focused anxiety, and self-dissatisfaction. Typical models tend to possess characteristics that are not typical of the average man or woman and their images are frequently further tweaked, airbrushed, or enhanced postproduction through the use of photo-editing tools such as Photoshop. Photoshop has altered our culture's perception of physical appearance and has been shown to heighten the negative effects of viewing perfect, unrealistic ideals. Viewing such images has been correlated with extreme dieting, body shame, increased cosmetic use and cosmetic surgery, low self-esteem, depression, anxiety, and eating disorders.

Some viewers are more affected than others. Some tend to internalize the ideals and strive to attain them while others do not. When standards are internalized, viewers who look different from the ideal tend to suffer from more detrimental effects. Internalization is increased by length of exposure, extent of identification with the model or product, and belief that the ideal is possible. Specifically, more lengthy exposure has been shown to lead to greater tendency to internalize the ideals, similarities in culture or race between the viewer and the model can increase internalization, and subtlety of modification can increase the effects. For example, research has demonstrated that when women are aware that images have been digitally altered, they suffer less negative internalization of the ideal. Knowing that the image is unrealistic, and thus less attainable, causes women to identify less with the image and to suffer less detrimental effects.

Chiara Rollero from the University of Turin found that women who viewed photos that were obviously retouched were less likely to internalize the ideals than the women who viewed photos that were more subtly retouched or not retouched. Thus, if the effects of Photoshop are obvious, women are less negatively affected by being exposed to them. Photos that were subtly retouched or not retouched caused greater internalization of beauty ideals, decreasing self-confidence and positive affect. Making the use of Photoshop explicit can decrease negative effects of viewing ideal images, but use of average women for media campaigns leads to even more positive effects. When women view ads that use models who are more representative of the average body, women tend to have increased body image and display more positive emotions.

The age of Photoshop increases the unattainability of media ideals. Even today's supermodels cannot achieve the perfection that is displayed on the pages of fashion magazines. Their images are altered until they defy the proportions of a viable woman. Since we are inundated with these images, they start to look familiar and real, but they are just outside the bounds of reality. To attempt to achieve this unrealistic extreme, many women, and increasing numbers of men, are pursuing cosmetic surgery. Breast implants, nose reshaping, eyelid shaping, and fat removal are becoming more and more common. Unfortunately, even beyond the normal risks associated with surgery, cosmetic surgery such as breast implants also run the risk of rupture, leakage, cancer, and autoimmune diseases.

Cultural Differences as Revealed by the Modeling Industry

Dr. Katherine Frith, Dr. Ping Shaw, and Dr. Hong Cheng examined fashion magazines from different cultures in an attempt to determine cross-cultural differences in attractiveness. They report that in the United States, femininity seems to be tied to being emotional, nurturing, passive, deferential, and physically attractive. In Asian cultures, the female ideal tends more toward modesty and goodness. In many cultures, advertisers use the female form and accentuate their sexuality and objectification to catch the attention of the consumer. France and the United States are particularly guilty of using the female form to sell products. Studies in China reveal more modest use of the feminine figure. These advertisers portray women in ways that preserve modesty, focus on the face, and use good taste to reflect the cultural ideals. When Chinese, Indian, or Japanese cultures do portray women as sexualized objects, the models used tend to be Caucasian, and they are portrayed in contexts uncommon to those of women of that culture. For example, lingerie ads in Asian magazines tend to be entirely made up of Caucasian models.

Researchers categorized the predominant types of beauty used in the modeling industry. The categories of the types of beauty were Classic, Feminine,

Exotic, Sensual, Cute, Girl-Next-Door, Sex kitten, and Trendy. The most common types used in U.S. fashion magazines today are Classic/Feminine, Exotic/Sensual, and Trendy. Classic/Feminine refers to women who present themselves as elegant, sophisticated, and fair. The Exotic/Sensual category refers to women who are presented as wearing revealing or tight clothing, and are posed in sexually attractive poses. Trendy models wear current clothing styles, wear oversized jewelry, and present rebellious attitudes in pose and appearance. Japanese magazines marketed toward teenaged girls use more Cute and Girl-Next-Door images. These models are presented as smiling or giggling and tend to appear youthful, cute, and modest. This is in direct opposition to U.S. magazines, which predominantly use defiant and independent expressions and images. Exposure to such models influences what individuals within the cultures may aspire to be and shapes how women view themselves.

To examine cross-cultural trends, these researchers used these categories to examine the content of popular fashion magazines from Taiwan, Singapore, and the United States. These three countries are relatively equal on standards of living yet are under varying cultural influences. Examining magazines in these cultures reveals the current preferences for beauty standards by culture. Dr. Frith and colleagues expected to find differences in the categories of fashion types presented between the cultures and also expected to find differences in poses used for Caucasian models versus Asian models due to different cultural expectations of beauty.

In the magazines in all three cultures examined, Caucasian models were used more often than any other race. In the magazines of all three cultures, a classic pose was used most often. After that, the choices differed by culture. In the United States, sexual poses were more common than in Taiwan ads or in magazines from Singapore. In Taiwan, Cute or Girl-Next-Door images were more often portrayed. In Singapore, models representing the trendy category were more common.

In the different magazines, the vast majority of the models were Caucasian and Chinese. When Dr. Frith and colleagues examined the beauty types by race, significant differences emerged. Caucasian women accounted for twice the number of models who could be categorized as sensual or sexual. Caucasian models were also more likely to be trendy and independent. On the other hand, Chinese models were more represented in the Cute or Girl-Next-Door categories. The final difference that emerged was that beauty products were more highly advertised in Singapore and Taiwan while the United States had more advertisements for fashion.

Dr. Frith and colleagues recognize the universal aspects of beauty. The classic beauty pose was the most common and the most similar even among these magazines that represented radically different cultural ideals. However, the sexualization of the female form is more rampant in the United States than in other parts of the world. Asian models are more likely to hold demure poses, regardless of the magazine examined and the Caucasian women are more likely to be sexualized. Furthermore, in the United States, the body tends to be the focus of attention, and in Taiwan and Singapore, the face was more central in the advertisements. Thus, in the ads from the United States, clothing and fashion are more common, whereas beauty products were more common in the other two cultures.

The Effects of Cultural Differences in Media Portrayal

Caucasian models tend to be overrepresented in magazines around the world. That means that many beauty advertisements highlight ideal characteristics of Caucasian women. Even models in magazines in other cultures have a larger than expected number of Caucasian models. If media exposure tends to cause viewers to internalize the cultural ideal, one might expect that this infusion of Caucasian models in advertisements may cause other races to chase Caucasian ideals. Alternatively, not being represented in the media, and thus not identifying with the idealized models, may eliminate the negative effects of the media. Peggy Chin Evans, from Michigan State University, and Allen R. McConnell, from Miami University in Ohio, examined the impact of beauty standards on different races of women. Since internalization of media ideals partially relies on self-identification with the model, groups that are not represented in the media could have a greater likelihood of maintaining higher self-esteem. Without an ideal to compare themselves to, individuals may engage in more realistic social comparison and compare themselves to others in their social groups. Comparing oneself to average women has been shown to lead to a higher self-esteem and less extreme personal ideals.

Black women, for example, are not as extremely affected by the thin ideal represented by the Caucasian models who are typically presented in media. It is possible that black women do not identify with the Caucasian models, and this may explain why they do not feel the same pressures to be abnormally thin and report fewer negative feelings about being overweight. They also hold less negative attitudes toward other women who are overweight. Overall, black women in the United States tend to have overall higher body satisfaction than white women.

To examine the extent of this effect, Dr. Evans and Dr. McConnell exposed women of different racial minorities within the United States to mainstream beauty standards and measured the resulting effects on their own self-evaluations. Since beauty advertisements typically contain images of Caucasian women, these researchers expected that, even though the groups studied are living as minority groups within a mainstream population, they would respond to standards of beauty differently. These researchers specifically focused on black, white, and Asian women.

As expected, black women did not find the mainstream beauty standards relevant to themselves and thus there was no impact on their self-evaluations after viewing the images. Viewing a thin Caucasian woman was simply not relevant to their self-perceptions and thus had no impact on their self-evaluations. Asian women, however, responded more similarly to Caucasian women and both groups, Asian and Caucasian, suffered from greater body dissatisfaction after viewing the mainstream standards. Since black women did not alter their self-perceptions after viewing a Caucasian ideal, black women tended to have higher self-esteem than either Asian or white women. Black women also demonstrated the lowest need to conform while Asian women had the highest need to conform, potentially explaining the differential effects of exposure to the media.

Self-esteem can be particularly affected if who one would like to be, or one's ideal self, does not match one's perception of one's actual traits. To increase self-esteem, individuals must alter their ideas about their ideal selves or make changes to their actual selves to become more similar to their ideal. When examined, the black women in this study had the closest match between their ideal selves and their actual selves, contributing to their higher self-esteem. The Asian women were found to have the largest gaps between who they are and who they would like to be. If Asian women are using Caucasian models as their ideal, it will be impossible for them to make who they actually are meet that ideal.

Speshal T. Walker also examined the impacts of white standards on the perceptions of black women in the United States. As noted, many ideals of beauty revolved around a Caucasian standard. Although there is a trend toward more diversity in advertising, even current black models tend to display features more typical of Caucasian women. This skewed representation may have a greater effect on self-perception than previous nondiversity. When black women were asked to rate their own level of attractiveness after viewing black models who had Caucasian-type features, self-esteem and body satisfaction were negatively influenced. Lighter skin was associated with higher ratings of self-attractiveness and those who perceived themselves as less attractive were more likely to take measures such as wearing hair weaves to make themselves more similar to the ideal.

Researchers were interested in exploring why Asian women are more susceptible to media pressures than black women. Part of the reason that Asian women may be more affected by beauty standards than black women is their underlying culture. The Asian culture emphasizes interdependence and conforming to expectations. Due to this, Asian women may be more likely to strive to conform to the dominant culture's norms more so than other cultural groups. In fact, Asian women report a stronger desire to conform to the mainstream cultural norms than even the women who were part of that culture (the white women). The result of such tendency is that Asian women are at higher risk for poor body image, high rates of body dissatisfaction, low self-esteem, and depression. There are preliminary findings that Hispanic women may be at a similar risk as the Asian women, as will be discussed later in this section. Since media overrepresents Caucasian models in magazines, on TV, and online all over the world, the impacts may be far reaching and how a group views the media may play a role in the magnitude of its effects.

In a previous study, Asian men and women rated white men and women as more attractive than Asian men and women. Since this group views white men and women as an ideal and since they are more likely to strive to conform, Asian men and women are more likely to take more extreme measures to alter their own physical appearances. This includes being more likely to purchase cosmetics, lotions, and other beauty products and to engage in cosmetic surgery, including eyelid surgery and nasal implants, to significantly alter their faces. Unlike any other cultural group, the surgery for Asian men and women alters the characteristics of their race rather than just reshaping a specific body part.

Since white models are most frequently represented throughout the world, Dr. Elizabeth Poloskov and Dr. Terence J. G. Tracey from Arizona State University extended the research and examined the impact of sociocultural standards of female beauty on body dissatisfaction of women of Mexican descent. Research has demonstrated that body dissatisfaction can lead to increased incidence of eating disorders, poor self-esteem, depression, and anxiety. The effects of body dissatisfaction have typically been studied in Caucasian women, but the onslaught of racially biased media predicts that other cultures will likely be affected as well. When Mexican American women were analyzed, the level of enculturation predicted how impacted they would be by the ideals portrayed in the media. The more the women were enmeshed in white society, the more influenced they were by the media ideals. Those more involved with their own culture were buffered from the negative effects of Caucasian ideals.

Some Caucasian ideals that may particularly affect minority groups living in the United States and Canada include traits such as hair type, complexion, and facial features, particularly for women. While some of the media ideals may be difficult for white women to achieve, the pressures to be young, thin, and blonde may be even less attainable to women from other cultures. Black women tend to strive to be lighter skinned and spend money to make their hair straight or wavy, regardless of their natural genetics and features. Over 80 percent of African American women in the United States straighten their hair. They also use chemicals to relax their curls, and conditioners, wigs, extensions, and bleaches to radically alter their natural features.

Furthermore, women are told that their bodies should be hairless. Time, money, pain, and effort are spent on carefully maintaining hair removal to be considered attractive and sexy. Again, over 80 percent of women engage in hair-removal practices and over 90 percent remove hair from their underarms and legs. It is also becoming more common for women to remove hair from their genitals despite the protective role of pubic hair. Hair removal can be painful, may lead to infection, and may lead to discomfort or irritation. Although hair removal seems like a personal choice, women who do not shave their armpits or legs are routinely met with expressions of disgust and condemnation from men and from other women, creating social pressures to conform to this beauty ideal.

Chasing beauty ideals also contributes to widespread cosmetic use. This practice, though expensive and time-consuming, is very common for women throughout the industrialized world. Wearing makeup increases the ratings of health, power, and capability. Women use makeup to feel attractive and professional. Darker-skinned women may feel pressures to use face-whitening products to achieve the media ideal of fair femininity. These products may

An African American woman has her hair straightened to achieve soft, wavy locks. Over 80 percent of African American women who are exposed to American media beauty ideals use chemicals, conditioners, wigs, extensions, or bleaches to alter their natural features. (Kouassi Gilbert Ambeu/Dreamstime)

include unsafe chemicals and cause skin irritations and damage. Still, nearly half of women from Hong Kong and a third of women from Indonesia, Malaysia, and Taiwan report that they routinely buy skin-lightening products. Interestingly, Caucasian women spend money darkening their skin through tanning and feel the results make them look healthier and wealthier.

POTENTIAL SIDE EFFECTS OF A PREOCCUPATION WITH BEAUTY

Impacts on Body Image

When one becomes hyperfocused on one's own appearance, one runs the risk of becoming self-conscious. An extreme focus on appearance can lead to difficulties in maintaining a positive body image and to self-objectification, or the tendency to judge and value oneself simply based on the body's appearance rather than on its effectiveness or function. Although self-awareness is a hallmark of being human, being hyperaware of one's own physical appearance is correlated with less positive feelings, higher rates of anxiety, and higher rates of depression. Clinical disorders such as eating disorders, sexual disorders, and mood disorders have been linked to being hyperaware of one's own appearance and hyperaware of the perceptions of others. Problems can especially emerge when one's perception of oneself does not match one's ideal. In the United States, consumers are continuously confronted with ideal body types via ubiquitous media sources, and this ideal can become more extreme and more difficult for an average individual to achieve.

The U.S. culture tends to focus on females' physical attributes and women are at a greater risk of defining their own worth based on their physical traits and their attractiveness to others. This slant can lead women to feel shame about their bodies, feel anxiety when in public, be hyperaware of their caloric intake, feel depressed, and feel less confident in sexual encounters, particularly for those women with lower interoceptive awareness (ability to monitor one's own internal states). Women with lower interoceptive awareness tend to place more emphasis on external cues and are more body conscious than those who have more internal awareness of their own body. Men, alternatively, are taught to focus on their physical effectiveness and tend to value strength, power, and dominance rather than individual features. Although this focus can also lead to problems, men tend to have fewer psychological problems as a result. Their focus is on function rather than on appearance. However, if men do become hyperfocused on not meeting a cultural ideal, they too can be negatively affected. Men may suffer from a lack of sexual interest, erectile dysfunction, lack of attraction to a partner, depression, anxiety, stress, and poor body image.

Empirical evidence shows that individuals in the United States have grown to associate a thin body with life success, particularly for women. Dr. Chin from Michigan State University demonstrated that exposure to a single photograph of a thin woman promoted social comparison and lowered a woman's self-satisfaction and optimism. In another psychological study, women who were exposed to thin, attractive models were more likely to have decreases in mood and body image. These effects on a woman's self-perceptions are likely a by-product of the tendency to self-compare. Those who have a greater tendency to engage in self-comparison tend to suffer the most. They are also the most likely to suffer from eating disorders, body dissatisfaction, and low self-esteem. Self-comparison tends to be heightened if the model looks similar to oneself, if the model is engaging in activities that one enjoys, if one identifies with the model in some way, or if one has increased duration of exposure. Models who do not have common characteristics with the viewer may be less influential. Furthermore, when women were provided with data that invalidated the connection between thinness and success, their levels of satisfaction and optimism improved.

Unfortunately, in many psychological research studies, when women are asked about their bodies, many discuss their bodies in terms of problem areas and fat deposits, and ugliness and age, rather than in terms of function and health. To examine how women view their own bodies, Ashley Mckay interviewed women from Canada and asked them about their race, class, and sexuality. Dr. Mckay found that women are aware that media pressures influence their body image perception. These women discussed the pressures to conform, and the negative impacts of the media on their emotions, self-perceptions, and confidence. Some seemed to be more impacted than others, with some actually expressing a feeling of threat from viewing the images.

Athletic women in Dr. Mckay's study reported difficulty maintaining feminine beauty standards while competing in male-dominated arenas. They reported struggling to find a balance between being competitive while not being threatening to the male ego. They discussed the struggle of deciding whether to be pretty or strong (or smart) and the social pressures that tell them that they cannot be both. Women endorsed feeling that they need to be attractive enough to garner attention and get opportunities but not too attractive or others will assume they are not very bright. Women also reported that the time needed to diet and exercise inhibits the time they can spend on other academic and professional goals. The women in the study also endorsed that there are time commitments that surround dressing professionally but remaining feminine. Hair, makeup, nails, dress, and shoes must be carefully orchestrated to balance societal expectations with personal goals.

The women acknowledged the importance of looking beyond the media to find their own self-worth. Each endorsed that their family and friends are instrumental in creating and maintaining healthy body images. Being accepted and reinforced for natural characteristics buffered their self-image in the face of the competing media images. The women in Dr. Mckay's study also mentioned that being part of multiple cultural groups also caused them to struggle. For example, living in an Indian household but attending a Western school caused conflicts in body image as the women attempted to conform to both norms. The women reported engaging in cosmetic use, dieting, and making careful clothing choices to avoid judgment from others and to increase their self-esteem. They skipped meals, altered their eating habits, and excessively exercised to feel better about themselves and to achieve more positive judgments from others.

The results of this study have been replicated and extended again and again. In many places around the world, women base their self-worth on their own level of attractiveness. Women are frequently scrutinized and assessed by other men and women, and may grow accustomed to being judged based on their physical appearance. Due to this awareness of being judged, if women feel unappreciated or unattractive in the eyes of others, it impacts their self-esteem and body image in ways unbeknownst to men. Unfortunately, preoccupation with physical appearance can be damaging to other realms of a woman's life as well. Simply making women aware of their own appearance, such as by asking them to try on a swimsuit, can alter their future behavior and performance. Researchers Barbara L. Fredrickson and Tomi Ann Roberts and colleagues were among the first to establish that after trying on a swimsuit, women are more likely to perform more poorly on a math problem and be more conscious of eating in front of others.

Francesca Guizzo and Mara Cadinu from the University of Padova reaffirmed that when women are reminded about their physical appearance, whether through becoming self-conscious or being noticed, commented on, or objectified by others, their cognitive performance decreases. Women tend to respond to objectifying gazes by taking up less space, talking less, becoming self-conscious, and performing more poorly on problem-solving tasks. They are more likely to fulfill cultural stereotypes and focus on their own bodies rather than on their cognitive abilities. In many cultures, it is common for women to be ogled, catcalled, and harassed, and thus there can be widespread implications for female performance and achievement. The implications for an individual woman are dependent on how much she internalizes the cultural beauty ideas promoted by the media and how well she buffers herself against the advances and judgments of others.

To minimize the negative effects of media ideals, avoiding exposure and not internalizing the media standards would best serve women. Women with strong friendships, supportive families, and low media exposure will likely conform less to beauty standards and will suffer fewer negative effects. For example, research has shown that overweight women who have not internalized the cultural standard of unnatural thinness were likely to remain healthy and confident. Overweight women who have internalized the ideal, however, were more likely to engage in disordered eating and be at risk for depression, anxiety, low self-esteem, and poor body image.

Prevalence of Eating Disorders

One of the largest growing areas of concern in the face of the current hyperthin media culture is the development of eating disorders. Based on the body focus predominant in industrialized cultures, being thin is a sign of success, and body size is commonly used to make superficial judgments about others. For those struggling in personal, academic, or professional aspects of their lives, being thin may be one way to bolster self-esteem. Being extremely underweight, though potentially less healthy, is typically preferred to being overweight because it conforms to the social ideal. As the ideal has become increasingly thin, dieting behavior has become more common and now typically begins in childhood for many girls. As new generations have been exposed to unrealistic body ideals portrayed in the media, more than half of young women report that they spend time dieting, engage in purging, use diet pills, and face psychological stress surrounding food and healthy eating. Currently around eight million individuals in the United States suffer from eating disorders and rates of anorexia and bulimia are becoming more pervasive for women and men.

Recently a phenomenon has emerged online of sites that are marketed toward inspiring people to be their best. Initially called "thinspiration," the sites claimed to help inspire individuals to lose weight to reach their ideal body size. However, many of the images portrayed on the sites were of unhealthy, extremely skinny women and the text supported behaviors that are indicative of anorexia or other eating disorders. "Fitspiration," a proclaimed healthier alternative to these inspiring websites, shifted the focus and encouraged viewers to be fit rather than thin. However, viewing perfect, toned, athletic, muscular bodies can still have an impact on one's own self-esteem. Given that fitness may look different from individual to individual, striving to look like someone else can be damaging to anyone's self-esteem. When women were asked to view photos of thin women in a psychological study, their ratings of their own body satisfaction plummeted. If the bodies were fit and muscular, the same effect was found. Thus, some of these websites that claim to support and inspire are actually endorsing unhealthy self-comparison, poor eating habits, and excessive exercise habits, and are contributing to increases in rates of mental health disorders such as eating disorders and depression.

Although there are multiple and individual causes for eating disorders, there is a link between incidence of eating disorders, such as anorexia, and self-perceived physical attractiveness. Clinical populations are particularly sensitive to decreased body image, have negative attitudes about their own attractiveness, and express increased preoccupation with physical appearance, particularly when confronted with media ideals. Those with anorexia are more likely to overestimate the importance of attractiveness and are sensitive to the possibility that being attractive will lead to positive social consequences. They are also less likely to rate themselves as attractive. This is one realm where preoccupation with physical appearance is detrimental to functioning.

Those struggling with eating disorders are more likely to have low body satisfaction, are much more likely to attempt to alter their appearances, and do much more checking of their appearance throughout the day. They are aware of and anxious about their own appearances. They rate attractiveness as more important than do healthy controls and believe attractiveness contributes more to success, competence, and other aspects of their lives such as interpersonal relationships, achievement, and happiness. They also overestimate the attractiveness of others (while underestimating their own attractiveness) and rate attractive individuals as more important than less attractive individuals.

Media exposure can exacerbate the symptoms of individuals who suffer from eating disorders and can even affect the healthy men and women who are typically used as controls in psychological research. For example, Cynthia R. Kalodner from West Virginia University examined the impact of media on men and women who do not suffer from eating disorders. She found that, for women, in particular, viewing thin models from media advertisements increased self-consciousness and body anxiety and decreased self-evaluations. In further studies, exposure to thin models for less than five minutes increased depression, guilt and shame, feelings of insecurity, and body dissatisfaction. These results were found with healthy women who do not already suffer from psychological disorders. Individuals with eating disorders are even more susceptible to such pressures, magnifying the negative effects. Men are not as affected by viewing thin models. Men actually experience a sense of increased body competence and feel stronger and better coordinated after viewing the images. Men may not internalize or self-compare to the media images because the ideal for men is to be muscular rather than thin. Men are more affected if they are asked to view images of muscular models rather than thin models.

Although women have traditionally been more likely to suffer from eating disorders and have been the subject of more study, there have been studies directed toward the effects of media on male body image and eating concerns. Researchers have found that viewing fitness magazines was correlated with eating issues and drive for muscularity for men. From the early 1970s, the size and muscularity of males portrayed in the media have steadily increased in television, movies, advertisements, print, and toys. Exposure to such potentially unrealistic ideals causes greater body dissatisfaction among males just as exposure to increasingly thin models affects women. Research suggests that the more exposure an individual has to cultural ideals, the more likely he or she will internalize the ideals and view them as personal goals.

Teresa L. Marino Carper and colleagues from the University of Central Florida examined the relationship between media ideals and body image among homosexual and heterosexual college males. Previous research has documented that exposure to mass media instigates preoccupation with physical traits for both men and women. Although the bulk of the research has focused on the effects of media portrayal of ideals on women, this research demonstrates that other groups are also affected. Specifically, Dr. Marino Carper and colleagues found that gay men were significantly more influenced by the media than straight men and were more likely to expose themselves to media ideals. Gay men were more concerned about being slender and were more dissatisfied and anxious about their bodies and more influenced by the cultural ideals portrayed in the media. They found that gay men rated physical attractiveness as more important and were more vulnerable to the effects of media ideals and much more likely to suffer from eating disorders.

Several other research studies examining the effects of media on psychological health has revealed that exposure to cultural ideals of increased muscularity for males and thinness for females has led to an overall increase in eating disorders for both genders. Gay men are particularly at risk for eating disorders and for body dissatisfaction. In one aspect, gay men are similar to women in that they are attempting to attract a male. Since males notably place more emphasis on physical appearance in attraction, gay males may feel more pressure to achieve physical ideals.

Physical Attractiveness and Depression

Physical attractiveness influences attitudes, attributions, and actions of the self and others. Attractive people are rated as physically and psychologically healthier than less attractive people. However, under inspection, Dr. McGovern, Dr. Neale, and Dr. Kendler from Case Western Reserve University and the Medical College of Virginia found that unattractive people did not display increased symptoms of depression, at least in twin populations. Depressed people do rate themselves as less attractive than people who do not suffer from depression. However, when rated by others, the same differences are not found. This means that feeling unattractive may contribute to depression more significantly than actually being unattractive. Similarly, being depressed may contribute to feeling unattractive, making one feel even more depressed, creating a negative feedback loop.

Steven W. Noles and colleagues from the Virginia Consortium for Professional Psychology and Old Dominion University examined the relationship between physical attractiveness and body image and depression. They also found that depressed participants were more likely to rate their own bodies as less attractive and were less satisfied with their own appearance than nondepressed individuals. When rated by others, they were not rated as less attractive, supporting the conclusion that depression is correlated with a distortion of body image. Interestingly, nondepressed individuals were found to rate themselves as more attractive than did objective observers. Research shows that body satisfaction is related to happiness and feeling unattractive is related to unhappiness, but it may also be the case that happiness leads to body satisfaction and unhappiness leads to feeling unattractive.

Dr. Gupta, Dr. Etcoff, and Dr. Jaeger also examined the effects of beauty on psychological well-being. Using a longitudinal study, these researchers found a relationship between objective physical attractiveness and psychological health. Attractive individuals in their study were less likely to experience distress and depression over the course of their life span. Dr. Gupta and colleagues found that facial attractiveness is positively linked with psychological well-being and lower rates of depression. The more attractive individuals were more assertive, had greater internal loci of control, and had higher confidence in social and professional environments. Furthermore, facial attractiveness was correlated with increased happiness and life satisfaction and higher satisfaction with romantic endeavors. These individuals also enjoyed slightly better mental health and less social anxiety. Furthermore, in an additional study, unattractive males had higher rates of self-reported depression. These findings conflict with the research presented at the beginning of this section, so attractiveness may not be the only key to happiness and psychological health.

Attractiveness and depression seem to have a complicated relationship. Low levels of attractiveness can influence depressed feelings and depressed feelings can influence one's perception of his or her own level of attractiveness, creating a negative feedback loop. Alternatively, higher levels of attractiveness can contribute to confidence, opportunities, successes, wealth, and status, and having such opportunities has been linked to lower rates of depression, regardless of actual appearance, creating a positive feedback loop. Thus, there are direct and indirect effects of attractiveness on the development of depression and of depression on self-ratings of attractiveness.

Physical Attractiveness and Anxiety

Research has demonstrated that being preoccupied with physical appearance can cause anxiety, particularly in women. This anxiety is heightened if one's physical appearance is not in line with societal ideals. Much research has demonstrated that exposure to the thin ideals in current U.S. media leads to at least short-term decreases in body image and increased anxiety, and women report feeling worse after viewing images of thin women than they do after viewing images of average women or other stimuli. Women who are dieting are particularly affected by viewing thin models. Women who are satisfied with their own weight were the least affected.

Emma Halliwell and Helga Dittmar from the University of Sussex examined the impact of the thinness of a model on anxiety levels of the viewer. In their study, the same model was used for all viewers to control for overall physical attractiveness of the model. Only the size of the models was digitally manipulated to create a thin or average image. Dr. Halliwell and Dr. Dittmar found that those participants who were exposed to thin models (rather than average-sized or no models) expressed greater anxiety, particularly for those women who internalized the thin ideal. Additionally, the advertisements using average models were rated as just as effective as the advertisements using thin models or no models. This is particularly revealing because many advertisers opt for thinner models in a mistaken assumption that the ad will be more effective. Thus, these modern advertisements not only increase the anxiety of the consumer but also do not effectively increase the success of the advertisements.

Additional research has revealed that age may also make an individual more or less susceptible to feeling anxiety after exposure to unachievable societal norms. Adolescents seem to be particularly at risk. As adolescents attempt to develop their identities, they are influenced by cultural guidelines and norms. These guidelines come from parents, friends, peers, role models, and the media. Because the media arguably contains individuals who are exemplars with ideal (or Photoshopped) characteristics, teens may struggle to meet the expectations and can develop social appearance anxiety. Many teens use magazines and the Internet to gather information about how they should look and how to attain such results, so these formats may be particularly damaging to self-esteem. Social appearance anxiety is particularly important to study because it has been linked with the development of eating disorders.

Jolien Trekels and Steven Eggermont interviewed adolescents to examine the impact of media exposure on social appearance anxiety. They found that the more time the adolescents were required to spend reading appearance-related material, the higher their rates of anxiety. They also found that the participants who associated attractiveness with increased rates of social reward were more motivated to internalize the attractiveness standards and experienced more anxiety when being viewed by others. Furthermore, such effects were found for both boys and girls. However, since there are more magazines that target girls and women, females tend to be affected more throughout the life span and tend to have higher rates of social anxiety related to their appearances.

CONCLUSION

We live in a media-saturated world that is full of exemplars of beautiful people. From childhood we are exposed to television shows, advertisements, fashion magazines, music videos, video games, billboards, and countless websites. Many parents make efforts to shield their children from the onslaught of aggression and sex inherent in advertising, but it is impossible to eliminate the exposure. As a result, media affects our expectations of others as well as of ourselves. This exposure can be positive, informative, and instructional, but it can also be psychologically damaging. The increased awareness of physical ideals can have many psychological impacts.

With the increase of the use of physical ideals in media, there has been a corresponding increase in struggles with self-consciousness, body image, eating disorders, depression, anxiety, and radical increases in elective cosmetic surgeries. Over the previous century, more and more individuals have become preoccupied with an unattainable ideal, and this preoccupation has had corresponding effects on psychological and physical disorders. Adolescents are particularly at risk for the negative psychological effects of exposure to media ideals. Maturity, experience, and more attainable role models can buffer the effects of media for adults. Adults who have personal role models, who accept themselves as they are, and who strive for health and moderation have better psychological outcomes than those who strive for perfection.

Part II
Beauty from Head to Toe

Part II of this text will more thoroughly examine the research on the individual features that contribute to one's overall attractiveness. Attractiveness is arguably more than just the sum of individual parts, but the parts do play a role. Here, physical characteristics, including those outlined in Chapter 1, will be examined individually. For each individual feature, the ideal may vary by sex, age, race, and culture. No one looks exactly the same, and minute differences in individual traits culminate to create one's overall physical appearance. Research that explains why some features are more attractive than others will be explored and cultural similarities and differences will be highlighted.

The following analysis examines the influence of hair placement, length, quality, and color; skin quality, texture, and tone; and body shape on the ratings of attractiveness. It examines specific facial, body, hormonal, and personality features that have been found to contribute to attractiveness. Finally, it includes sidebars that go into more depth regarding specific cultural exceptions for a variety of these features.

HEAD, FACIAL, AND BODY HAIR

The presence, placement, and quality of head and body hair differentially influence male and female ratings of attractiveness. Hair length, color, and quality can serve as indicators for female age and fertility, while hair placement and thickness provide cues of masculinity.

Hair length, color, and quality are significantly correlated with female attractiveness. Long healthy hair indicates good nutrition and overall good health, and provides some indication of age. Cross-culturally, younger women tend to grow their hair longer than older women, who tend to maintain their hair at shorter lengths. Thus, hair length is an initial advertisement for age, which signals the level of reproductive value. Hair also grows fastest in women who are at peak fertility, naturally giving them longer, healthier locks. Additionally, lighter hair colors are associated with youth and vitality. Hair tends to darken with age, making hair color one cue that can be used to determine approximate age. Finally, hair quality is used an as indicator of health, nutrition, and level of personal care across cultures and thus is particularly associated with the ratings of female attractiveness. Unmanaged, unwashed, or mangy hair significantly lowers attractiveness ratings regardless of other features.

Hair removal can also be associated with increased attractiveness. Many women and men around the world remove hair from different body parts and sculpt and tweeze the remaining hair to increase perceived levels of attractiveness. Women commonly remove hair from the body to increase the appearance of smooth, youthful skin, and men shape head, body, and facial hair to increase signs of masculinity and increase the appearance of overall symmetry.

Hair Length and Femininity

In the United States and other industrialized cultures, longer hair on the head tends to be associated with increased ratings of femininity. At points throughout history, long hair was fashionable for males as well, but overall, longer hair tends to be a cross-culturally feminine trait, likely due to the influence of estrogen on hair growth. This widespread trend for women to have longer hair illustrates the underlying hormonal functioning and enhances sexual dimorphism between the genders. Based on the tendency for women in most cultures to wear their hair longer than males, researchers were interested in whether long hair actually adds to female attractiveness from the males' perspective. Most research on this topic endorses that hair length can impact attractiveness ratings, with longer hair being rated as more attractive than shorter hair.

To illustrate, Tamas Bereczkei and Norbert Mesko from the University of Pecs in Hungary had 30 men rate female faces for level of attractiveness. Throughout the process, the 10 most and the 10 least attractive faces were identified and then given varying hairstyles via a computer program. When a new group of 80 males rated the faces, previously unattractive faces that were given medium or long hairstyles were rated as more attractive than they were in their initial condition. Long hair enhanced the ratings of attractiveness even for the faces that were initially rated as the least attractive. Since hormones control hair growth, hair length may be an honest cue of underlying hormonal functioning. Hair length impacted facets beyond physical assessments as well. In Dr. Bereczkei's study, hair length was also found to impact personality judgments. Longer hair was associated with maturity, intelligence, femininity, and health. Although typical hair length may be a by-product of social norms, there is consistency among American and European nations with longer hair leading to an increase in positive social judgments.

Although this research was done in the United States and Europe, there is complementary evidence of cross-cultural consistency. In many cultures, younger women tend to wear their hair longer and then to cut it shorter with age. Longer hair becomes an initial visual cue for possible fertility, sexual interest, and availability. Furthermore, in many cultures, hair is more typically worn up or covered after marriage when it is no longer needed to attract a male, providing additional information about the woman's availability and interest. Most women likely are not conscious of the signals they are providing with something as simple and personal as the length of their hair, but it nevertheless seems to function as a universal signal in the social environment.

Since hair length has been shown to increase the ratings of attractiveness, Gizella Baktay-Korsos from the Etovos Lorand University in Hungary was interested in examining whether there were any other benefits of having longer hair. Her research was based on the findings that longer hair in women is associated with increased perceptions of power, sexuality, and success. Dr. Baktay-Korsos questioned whether or not there was any truth in these perceived associations.

Dr. Baktay-Korsos started her research with girls and examined their experiences throughout school. She found that, from childhood, girls with longer hair actually had more friends, were rated as more attractive, were more popular, and were more successful when compared to either girls with short hair or boys. There could be several explanations for this association. One explanation could be that the genetic quality that underlies the growth of strong, healthy hair has benefits in other realms as well, such as physical and mental health and developmental stability. Additionally, since girls with long hair are perceived to be more attractive, others may have other positive expectations and evaluations of them as well, as a mere by-product of the halo effect. Dr. Baktay-Korsos also suggests that caring for long hair takes more time, so girls with long hair may have parents who are more invested. Parental investment and attention are highly correlated with successful development; therefore, those girls with a supportive home environment develop a stronger sense of self, have increased levels of confidence, and develop better social, cognitive, and emotional skills.

There are exceptions to this rule of hair length contributing to attractiveness. For the average women, longer hair does increase their average ratings of attractiveness. For the most physically attractive women, however, shorter hair has been used to increase their level of physical appeal. Some top models in the United States, for example, have very short hair. This lack of hair contributes to the ratings of attractiveness because their facial features are highly attractive. Without hair, the viewer's attention goes to the features of the face, so for these women, hair would just distract from their facial beauty. Short hair on these models increases their ratings of attractiveness, perceived power, and confidence. While for the average women longer hair increases the ratings of attractiveness, there are some exceptions for the most beautiful women.

Ultimately, long hair emerges as a trait that increases the ratings of attractiveness for females, cross-culturally. Long hair signals youth, fertility, and health and is correlated with greater success, higher self-confidence, and stronger social skills. Thus, research demonstrates that hair length is one way to increase perceived levels of attractiveness, at least for women.

Hair Color

In addition to hair length, hair color also has an effect on the ratings of attractiveness, although the ideal color varies by gender. Although there are certainly individual differences in preference, upon study, an overall preference for women with blonde hair tends to emerge, particularly in U.S. and European samples. Blonde hair is correlated with youth and hair tends to darken with age. Since youth is a key factor in the ratings of attractiveness for women, the youthful hue of blonde hair gives a woman an advantage over those with other hair shades. Adult women with blonde locks are likely to be rated as younger than they are when they are shown with darker hair. Natural blondes also typically have slightly higher estrogen levels than individuals with darker hair colors and have more stereotypically feminine facial features. The opposite is true for men. Since attractive men tend to be those who are more distinguished and successful, darker hair, belying maturity and strength, tends to be more attractive to women. The image of the tall, dark, and handsome man can be brought into play when considering an attractive male.

This area of research is muddled by individual preferences and the bias of dating experience, but overall, there are trends for a preference of blonde hair in women and darker hair in men. Most research studies find that lighter haired women have a physical advantage over darker haired women. Beginning in 1978, Dr. Feinman and Dr. Gill conducted a research study that asked men to rate the attractiveness of female photographs. They found that men rated blonde women as significantly more attractive than women with any other hair color. Through this research, Dr. Feinman and Dr. Sill empirically demonstrated that the average male has a significant preference for light-haired women. In a follow-up study looking at men's attractiveness, they found that women demonstrate a similar preference for dark-haired men.

Since self-report data are difficult to substantiate, Melissa Rich and Thomas Cash from Old Dominion University designed a study where they analyzed actual behavior to assess the impact of hair color on perceived attractiveness. In their research, they analyzed *Playboy* centerfolds to discover the level of inclusion of women with different hair colors. They found that, even though blonde hair is relatively rare in the overall population, it is highly represented in these men's magazines. Based on an analysis of a variety of magazines over a 40-year span, Dr. Rich and Dr. Cash found that blondes are more prevalent in *Playboy* centerfolds than in any women's magazines. Blondes make up approximately a quarter of the white female population, but in *Playboy* they accounted for almost half of the photographs. They found that blondes also tend to be overrepresented in movies and television in the role of pure and good characters and are the typical caricatures of the most beautiful women in men's magazines and Hollywood films. Blonde women are more likely to be cast as innocent princesses, angels, saints, goddesses, and fairy godmothers. Men rate blonde women as more attractive, feminine, emotional, and pleasurable. Evil characters like witches, devils, and criminals are more likely to have dark hair. Ultimately, from a cultural perspective, attractive women are fair and open while attractive men are tall, dark, handsome, and mysterious.

Later research empirically demonstrated that there is more variation in male preferences than in female preferences. Dark-haired males, for example, sometimes endorse a preference for dark-headed females. Alternatively, blond males were as equally likely to express a preference for blondes as they were for brunettes. However, a significant number of blonde, brunette, and red-haired women endorsed a preference for brunette males.

Ultimately, in psychological and self-report studies, men do endorse that brunette women possess positive qualities, but the term "beautiful"' tends to be associated with blonde women. Given such findings and given the purported importance of being beautiful, the tendency for women to dye and maintain lighter hair color is explained. However, although blonde women are endorsed as the prettiest, further research demonstrated that they are less likely to be judged as competent or successful. Thus, in modern society, women must find a balance between perceived beauty and perceived competence. If a woman's sole goal is to attract a man, blonde hair dye could be an effective step in that process. However, long-term success, not only in personal relationships but also in the attainment of career and personal goals, undoubtedly muddies the best strategy.

To make matters even less black and white (or blonde and brunette), Viren Swami and colleagues from the University of Westminster and the University College of London report results that contradict the typical findings. Dr. Swami self-reportedly recognizes that studies in North America overwhelmingly find that blondes are rated as more attractive than brunettes. However, in their research, the participants, particularly the women, rated brunette women as more attractive and more fertile. This is in direct contradiction to typical results. The researchers explain their findings by pointing to current fashion and style icons who have dark hair. Culture, social norms, and personal experiences likely alter what is considered most attractive; thus, it is difficult to find a universal preference for hair color that is untainted by socially constructed images of what a man or a woman is supposed to look like.

Beards and Male Body Hair

Features that enhance the distinction between the sexes, or those that increase the level of sexual dimorphism, tend to enhance the level of perceived attractiveness. Thus, it follows that male-typical beards and body hair should be rated as more attractive to women because they increase the distinction between males and females. Darwin considered beards an adornment of the human male. Similar to long hair in females, beards are regulated by genetics and enhanced by hormone exposure, so they should provide an honest signal of underlying physiological and genetic health. Full beards also can serve to enhance the perceived symmetry of the male face, if well maintained, increasing the ratings of attractiveness, particularly for those who are otherwise asymmetrical.

Barnaby Dixson and Markus Rantala from the University of Queensland and the University of Turku, respectively, examined the impact of beard and body hair on attractiveness ratings. Since facial and body hair is a defining feature between men and women, it was expected that women would find such traits attractive. Unsurprisingly, Dr. Dixson and Dr. Rantala did find that women rated men with facial hair as more attractive than those without facial hair. Specifically, Dr. Dixson and Dr. Rantala found that women preferred men with continuously distributed facial hair covering the lower jaw, mustache area, and cheeks to men who were clean-shaven or who had patchy facial hair. However, men whose bodies were clean-shaven were rated as more attractive than those with body hair. The only exception was those photos of men with light hair around the sternum, pectoral region, and areolae. Their research revealed no difference in preference when the stage of menstrual cycle was addressed, although given the difficulty in predicting fertility, especially using an online survey format, further research may be required in this domain.

Although a preference for beards and body hair would be expected from an evolutionary perspective, this area of research has had mixed results. Dr. Dixson and Dr. Rantala did not find that the level of attractiveness increased with body hair. However, other research shows marked increases in the level of attractiveness with increased body hair in other cultures. For example, women from the United Kingdom tend to prefer hairy chests, while women from China and many other industrialized nations express a preference for little to no chest hair. Furthermore, in an additional study, Dr. Dixson and colleagues demonstrated that women from other ethnic groups (Europeans living in New Zealand and Polynesians from Samoa) do not rate bearded men as more attractive than their clean-shaven counterparts.

Attraction to facial or body hair may vary with a woman's ultimate goals. If women are looking for a committed, caring partner, they may steer away from the more masculine men with hairy bodies and faces. Clean-shaven men tend to be rated as younger, more innocent, and more sociable. Since facial and body hair tends to be correlated with the ratings of masculinity and testosterone levels, women may forgo this more masculine feature for a less masculine but more committed, supportive, and sociable partner. Research does show that more masculine men are more likely to cheat on their partners and more likely to engage in short-term rather than long-term relationships, so facial and body hair may provide a cue for likely future behavior. Depending on the goals of the female, such characteristics may come across as either more attractive or less attractive.

Since the presence of facial hair is so variable between cultures and throughout time, researchers next considered whether facial hair might have another underlying purpose rather than to attract women. In their study, Dr. Dixson and colleagues debated whether facial hair might serve as a signal to other men rather than as a signal to women. They presented men and women with photographs of men who were either clean-shaven or bearded. The men in the photographs displayed a neutral facial expression, a happy facial expression, or an angry facial expression. Men and women then rated the faces for a variety of traits, including attractiveness and aggressiveness.

Findings revealed that women rated the men with clean-shaven happy expressions as more attractive than any of the other photos while men judged the bearded angry faces as much more aggressive looking than the clean-shaven men who were making angry faces. Thus, the power of the beard may function to scare or intimidate other men, rather than solely as a signal of attractiveness for women. Beards tend to make the jaw look larger, making the individual appear more confident in an altercation.

The results from Dr. Dixson's study agree with the results from other researchers who found that bearded faces were rated as more emotionally mature, sincere, masculine, self-confident, and brave than clean-shaven faces. Beards may, however, make the man appear less warm or cooperative, decreasing the effects of the positive qualities on female ratings. Level of grooming can also impact female ratings. A groomed beard can enhance perceived age, power, and confidence, but a poorly groomed beard can signal low socioeconomic status and social status.

Thus, it seems that men are at the mercy of a delicate balance as well. If a man maintains a full beard, he is likely more intimidating to other men and will be perceived as more masculine, mature, and powerful. However, such benefits may come at the mercy of being perceived as less supportive, cooperative, and caring by women. Ultimately, this demonstrates that power and status are possibly more attractive to women than actual physical features, so a beard may be a risk a man is willing to take.

Baldness and Masculinity

Baldness is a trait that affects males more frequently than females. Although most men may not desire to lose their hair, baldness has been found to have some redeeming qualities in the arena of attraction. Baldheaded men (both those who are naturally bald and those who shave) are rated as more dominant, stronger, taller, and more assertive, and as having more leadership potential than those with a full head of hair or those with thinning hair. These findings held consistent when photographs of men with full heads of hair were digitally altered to make them bald. The men who were digitally altered to look bald in the photos were rated as stronger, taller, and more dominant than the same men with hair. To see if these same perceptions held in the absence of a visible stimulus, researchers gave written descriptions to women, and those men who were described as having a shaved head was rated as significantly more masculine, more dominant, stronger, and with higher leadership potential.

A man with a bald or shaved head tends to be perceived as older, more distinguished, and more confident, and is less likely trying to hide something. When looking at a man, women tend to notice his hair first. In the absence of hair, a woman's attention goes to his eyes. This may create the perception of a quicker connection, increasing the levels of attraction. Furthermore, the characteristics of the groups of men who typically shave their heads, such as those in the military, law enforcement, and gangs, may increase baldness's association with power and dominance. Men with bald or shaved heads are seen as bolder, more masculine, more intelligent, more mature, and more honest.

Since most men lose at least some hair by age 35, the attractiveness potential of a shaved head is good news for the male ego. Thinning hair is rated quite low on attractiveness scales, so shaving one's head may be a recommended course of action for those men if they are interested in increasing their ratings of attractiveness. Michael S. Wogalter and Judith A. Hosie from Rensselaer Polytechnic Institute and the University of Aberdeen, respectively, examined the placement and distribution of hair for males to analyze how it impacts attractiveness. They found that those with less cranial hair were rated as older, and those without any cranial hair were rated as more attractive and more intelligent. So, although balding males appear older, they also are rated as more intelligent and more masculine.

Men's manipulation of facial, cranial, and body hair can communicate information to others. Harnessing the power of a beard can influence facial attractiveness and facial symmetry. Beards can communicate increased levels of masculinity to women and can intimidate other men. Furthermore, manipulation of cranial hair can send signals about the levels of intelligence, masculinity, and maturity as well.

Impact of Body Hair Removal for Women

While men can harness the power of hair to signal virility and strength, women in the Americas, across parts of Europe, and throughout Australia regularly remove hair from parts of their bodies in an effort to enhance their own attractiveness. If body hair signifies masculinity, then removal of hair arguably increases femininity. Most women throughout these regions shave or wax their legs and underarm areas and almost half remove portions of their pubic hair. Many sculpt eyebrows, and the hair on their arms, stomach, face, and toes. One reason behind such removal may be to enhance the smoothness of the skin, a known determinant of the level of attractiveness. Interestingly, since body hair is a sign of maturity, it seems counterintuitive to remove it. However, lack of body hair may also contribute to appearance of youth and further serve to distinguish the feminine body from the hairy masculine form. Because body hair removal has become such a social norm, women who do not remove body hair are consistently rated as less feminine, less sexually attractive, less sociable and intelligent, and more masculine.

Marika Tiggemann and Suzanna Hodgson from Flinders University in South Australia surveyed 235 female university students about hair removal practices. They found that nearly 100 percent of the women removed leg and underarm hair on a regular basis. Out of those who reported not removing hair (<3 percent), most reported having done so in the past or on an irregular basis. Seventy-five percent endorsed removing hair from their bikini line and over 60 percent reported shaving or waxing their pubic hair. When asked about their motivations, the women in this study reported that removal of hair made them feel more feminine, more confident, and more attractive. They also reported liking the feeling of smooth legs and feeling more sexually attractive. The women who removed leg, underarm, and pubic hair were also more likely to read fashion magazines and to watch programs like *Sex in the City* and *Big Brother*, demonstrating that they may be more attune to cultural norms.

Since body hair, specifically pubic hair, provides a function for health such as protecting the vulva from bacteria, and removal causes more health risks like infections, irritation, and spread of sexually transmitted diseases like herpes, it is surprising that its removal is so widespread. The origin of hair removal seems to be via a social or cultural mechanism rather that a biological one. Those who watch programs that glamorize female sexuality and promote hair removal to increase sexual appeal and those who read fashion magazines that instruct what an ideal woman's body is supposed to look like are more likely to engage in such practices. The women who shave their pubic hair are also more likely to be in intimate relationships than those who do not, illustrating the social pressure that may be involved in such a practice. Women are less likely to shave their pubic hair for themselves but will do so for a partner. As a by-product, the hair removal business is currently a multimillion-dollar industry.

The removal of body hair on women has become such a cultural norm that having hair, such as under the arms, has actually been shown to elicit a shock and disgust response from men and women. Exposure to fashion magazines and other forms of media has caused women to internalize the hairlessness ideal in many parts of the world. Women attempt to alter their bodies to match the cultural ideal to increase their attractiveness. In a previous study, Marika Tiggemann found that women, when presented with images of hairy female bodies, tend to have a more reactive disgust response than do men, but that both males and females react with disgust when confronted with female body hair. Since hair removal is a personal choice, some resist the cultural pressures to shave their legs and armpits. However, surprise and derision from others may pressure such individuals to hide their hair or conform to shaving practices.

Eyebrows

Eyebrows constitute another sexually dimorphic feature that typically differs between the sexes. Although both sexes have eyebrows, the ideal eyebrow thickness and placement are distinctly different between men and women.

Ideal male eyebrows tend to be thicker and bushier than female eyebrows, and the ideal distance from the brow to the eye is less in men than in women. Women who are rated as highly attractive tend to have higher, thinner brows than those of the typical man. Because of this, women tend to groom their brows, specifically the underside, to increase the distance between the eye and the bottom edge of the eyebrow. This pruning accentuates the distance between the eye and the brow, which serves to make their faces appear more feminine. Plucking also thins the brow, decreasing the contrast between the brow and the skin, further accentuating femininity. Since men tend to naturally have bushier, thicker brows, a strong contrast between the brow and the face is seen as a more masculine trait.

As women age, their brows tend to naturally thin. Thus, ideal eyebrow thickness is a balancing act for women. Fuller eyebrows on women signal youth, which contributes to attractiveness, but fuller eyebrows are also less feminine. So, the thickness of the eyebrows creates a tug of war between youth and femininity for women, which directly changes female grooming behavior over the life span. Younger women frequently pluck their brows to increase their level of femininity, but this behavior has the contrasting effect of making them look older. Older women are less likely to pluck their brows and may even start filling them in with cosmetics to achieve a more youthful appearance, even though this practice has the effect of decreasing their appearance of femininity. Essentially, younger women typically opt for femininity over youth while older women tend to opt for youth over femininity.

The height of the eyebrow on the face also influences the ratings of attractiveness. A higher brow is correlated with youth and submissiveness. Higher brows communicate openness, innocence, and passiveness, which are traits that are typically considered more attractive for females. Lowered eyebrows are indicative of dominance and aggression. If we put this in terms of typical facial expressions, a furrowed brow is more threatening and intimidating than a brow raised in surprise or fear. Thus, the high brow of an attractive female face may communicate youth and indicate a level of submissiveness that adds to her attractiveness. A thicker brow communicates dominance and strength and someone who is more difficult to control. This may be beneficial for a female's success in business or in other positions of power but is generally considered less traditionally attractive. A bushier furrowed brow on a male face, however, is considered to be attractive. A strong brow communicates authority, dominance, and control, prized masculine traits. A sturdy brow line makes a male look more powerful, masculine, and attractive, particularly in a woman's most fertile phase. A more narrow, high brow on a man, alternatively, increases his perceived level of femininity.

Although a thick, masculine brow has traditionally been considered to be less attractive for women, this fashion is at least partially affected by societal trends. A strong, thick brow line is currently a fad for women, and many young women are using cosmetics to enhance rather than decrease their brows. Thus, research demonstrating the ideal brow may not hold up to changing societal pressures. The most attractive brow height also varies depending on the age of the rater.

Media representations of the ideal eyebrow have changed over the previous decades, so different age groups have grown up with different societal ideals. For example, raters under the age of 30 rated lower eyebrows as the most attractive for women. These individuals likely belong to the current trend preferring thicker, lower brows for women. Raters over the age of 50 had the opposite preference and rated women with higher, arched eyebrows as the most attractive. This change in preference could be due to cultural changes and changes in fads in the media and entertainment over time. This difference in findings brings up an important issue in attractiveness research. Although there may be cultural universals, there is also room for cultural effects. When identifying an ideal, it may be necessary to study it not only cross-culturally but also across age groups. Fads, styles, and trends may affect entire generations and may vary across time.

SKIN

Human skin holds the distinction of being the largest organ in the human body. Skin plays the vital biological function of protecting us from the outside world, sensing the external environment, and maintaining internal homeostasis. Every person's skin creates a unique physical appearance, and among individuals, skin varies by complexion, color, and texture resulting from our individual genetic makeup and our developmental environment. The skin that covers our body creates the physical display that presents us to the external world. Skin's complexion, color, and texture vary on the level of attractiveness, typically corresponding to underlying physical health and self-care. The next three sections will explore the different qualities of skin: complexion, color, and texture.

Complexion

One's complexion is composed of the skin's smoothness and evenness of tone. Unsurprisingly, smooth skin is overwhelmingly considered to be more attractive than rough or scarred skin. Although many adolescents struggle with their complexions during and following puberty, most appearance-conscious adults work to present a clear and smooth complexion to the world. Since the skin of the face and arms is usually the most

Ritual Scarring in Ethiopia, Papua New Guinea, and Australia

Research shows that smooth, even skin is rated as most attractive around the world. Thus, the practice of ritual scarring creates an interesting exception for attractiveness research. Ritual scarring consists of rubbing small incisions with ash to enhance scarification, creating intricate symmetrical patterns on the face and body. This practice has been common across many small-scale societies across Southeast Asia and Africa, in groups such as the Karo tribe of Ethiopia, the Sepik region of Papua New Guinea, and the Aboriginal culture of Australia. Individuals from these cultures create intricate patterns on their faces and bodies for a variety of purposes.

In the Karo tribe of Ethiopia, ritual scarring is used to communicate strength, power, and beauty. Men use the scars to indicate status and power as well as to represent past achievements in warfare. The pattern, number, and types of scars communicate accomplishments such as success in hunting or victory in battles against other tribal groups. Women of the Karo tribe also scar their bodies, creating symmetrical patterns that indicate sexual maturity and enhancing sexual attractiveness. This scarification demonstrates maturity and sexual readiness, and increases the ratings of sexual appeal. Women with scarification are rated as the most attractive and the most alluring, and have the highest mate value.

In Papua New Guinea, scarification is similarly used to designate maturity, social status, and strength. The trials associated with earning and creating the scars are painful and demanding and mark the transition to maturity. Only those strong enough to endure the trials and the scarring process earn the privileges of status within the group. In Australia, similarly, scars denote maturity and carry with them the privilege of participating in the adult rituals of the Aboriginals. The scars display an outward signal of status and endurance. In this manner, ritual scarring increases attractiveness because it signals maturity, success, and strength.

Ritual scarring is used for many purposes such as to enhance beauty, communicate success in warfare, commemorate rites of passage, demonstrate solidarity with a particular group, demonstrate courage and endurance, mark physical maturity, and signal sexual readiness. On women, the patterns, placement, and number of scars contribute to the ratings of sexual attractiveness and mate quality. On men, such scars indicate strength, success in warfare, and high social status. In these societies, the ritual scarring techniques give clear outward signs of wealth, power, and enhanced symmetry, thus explaining why they contribute to the ratings of attractiveness.

visible to others, it tends to be cared for the most diligently and is used the most for information about the age and health of others. Clear, smooth, unblemished skin indicates maturity, health, and good grooming habits and contributes to attractiveness for both men and women. A clear complexion indicates lack of disease, parasites, or infections. Clear skin boasts of strong immune system functioning. Clear skin also provides some initial information about age because prepubescent individuals tend to struggle more with acne and older individuals have more blemishing wrinkles that come with advancing age. Reproductively viable individuals tend to have the healthiest, clearest complexions. For women, smooth, wrinkle-free skin is associated with low androgens and high estrogen. Accordingly, men rate smooth, clear, rosy complexions as the most attractive, and these traits correlate to healthy hormone levels, fertile youth, and healthy immune systems.

Acne, bumps, discolorations, infections, tumors, lesions, and other skin problems that disrupt the complexion decrease attractiveness in a radical fashion. The presence of such symptoms significantly decreases the ratings of attractiveness and intentions to pursue relationships. Presence of such problems can indicate disease and poor health, and individuals with skin problems tend to be rated as the least attractive cross-culturally, for both genders.

An exception to the preference for a smooth, healthy complexion is subtle signs of scarring on the male face. Researchers from the University of Liverpool and the University of Stirling engaged in a study with over 200 individuals to examine the impact of facial scarring. They discovered that subtle facial scarring actually increased the attractiveness of males. Since scars indicate past trauma or hostile engagements, they may be used by the rater as an indicator of dominance, extreme masculinity, and previous risk-taking behavior. Although such traits may not be an indication of long-term commitment, they could indicate enhanced traits of masculinity, dominance, and strength that tend to be preferred for short-term affairs. Thus, such indicators of past altercations can actually increase a man's level of attractiveness. Too much scarring, however, or scarring noticeably resulting from acne typically decreases the ratings of attractiveness.

Based on the high importance placed on complexion, it is no wonder that the cosmetic industry is flooded with hundreds of lotions and creams that claim to smooth wrinkles, decrease blemishes and scars, and increase the skin's suppleness; and with a wide variety of foundations and concealers marketed to women to help them hide flaws in their complexions. The curious thing is that males do not tend to partake in these products. Their skin is a similar advertisement for women and the appearance of smooth skin radically increases attractiveness ratings, yet cultural pressures create a taboo against men using these feminine products. Although women may not worry as much about their partners' ages, they do like clear, healthy skin. The onset of the metrosexual in the last decade may be demonstrating a change in male behavior in the United States. More males are using moisturizers, seeking skin treatments, frequenting spas, and engaging in cosmetic surgery to enhance their own levels of attractiveness. Furthermore, cosmetic companies have started marketing products specifically to men, complete with more masculine containers and color schemes.

Skin Color

Skin color is a component of one's complexion. In the quest to identify whether there is a most attractive skin color, researchers have faced conflicting results. Expressed preference for different skin colors varies by region, culture, and individual. Much early research demonstrated that lighter skin was typically found to be more attractive and more sought after across cultures. However, this preference is likely socially constructed. Historically, darker-skinned individuals have been commonly oppressed and lighter-skinned individuals have held the power and wealth. Furthermore, poorer individuals historically spent more time working outdoors in the sun and had darker skin as a result. Thus, it naturally follows that, in the past, individuals with lighter skin would be selected as more attractive since lighter skin was typically associated with the higher socioeconomic status.

In accordance with this finding, Viren Swami and colleagues from the University of Westminster and the University College of London found that participants from universities in London rated lighter-toned women as more attractive than tanned or darker-toned women. However, Bernard Fink and Karl Grammer form Ludwig-Boltzmann-Institute for Urban Ethology and Randy Thornhill from the University of New Mexico found that Caucasian males from the University of Vienna rated darker skin as more attractive, demonstrating that a preference for lighter skin is by no means universal or innate. Dr. Fink and colleagues pointed to social constructs to explain their findings. For the group of individuals being studied, tanned skin was likely used as a symbol of the amount of leisure time spent outdoors. In their study, pale skin would have likely indicated increased time spent working indoors, likely at an office or a factory, with little time for leisure. Thus, darker skin in this study communicated a healthy, relaxed, leisurely lifestyle, which added to perceived level of attractiveness of the individuals being rated.

Skin Bleaching in Tanzania

Ideal skin color within a society depends upon socioeconomic factors. In many societies, lighter skin has historically been correlated with wealth, status, and power, so many cultures still demonstrate a preference for lighter skin. In Tanzania, for example, lighter skin is rated as more attractive and native Tanzanians commonly engage in skin-bleaching practices to achieve lighter skin tones. Skin bleaching involves using lotions to lighten the pigment of the skin. These products include bleaching agents, and may also be mixed with detergent or battery acid to maximize their effectiveness.

Lighter skinned individuals in Tanzania tend to have higher social status, access to more resources and better jobs, receive higher bride prices, and are rated as more desirable and more beautiful. Unfortunately, despite the social gains, the processes and chemicals used to achieve skin lightening can cause acute health problems. Side effects of bleaching include development of blotchy uneven skin tones, rashes, peeling skin, skin and liver cancer, heart problems, and diabetes. These products have also been linked to organ failure and infertility.

Kelly M. Lewis from Georgia State University and colleagues examined bleaching practices throughout Tanzania to determine why skin bleaching is so common. Dr. Lewis identified that those individuals most likely to bleach their skin include older, wealthier women who have been exposed to Westernized standards of beauty. Dr. Lewis explains the trend using historical and cultural influences, as well as modern social pressures. African history is marred by periods of slavery and colonization, by light-skinned, wealthy Europeans. Early attempts to emulate these dominant groups led Africans to color their skin with henna or natural whitening pastes. There was and is a divide in power, resources, and status between lighter- and darker-skinned individuals. Even in their own communities following their independence from German reign, the lighter-skinned individuals were more likely to gain power and to be seen as more attractive and superior. Even recently, billboard advertising was found to depict lighter-skinned individuals as being more professional and successful and darker-skinned individuals were depicted as manual laborers or unemployed.

Tanzania is not alone. Skin-whitening products are ubiquitous in Japanese cosmetics advertisements. In Japan, millions of dollars are spent annually on skin-whitening products. Whiteness is associated with beauty, and Japanese women report purchasing whitening skin products, sunscreens, hats, parasols, sunglasses, and vitamins, and report staying at home or in the shade when the sun is strong.

For skin color, social and cultural constructs likely contribute to preferences. Thus, there is no one most attractive skin color. Attractiveness of skin color seems to be tied to who has the wealth and power within a society. If the wealthy spend most of their time indoors protected from the sun and the poor must work in the hot fields, then lighter skin is prized. Alternatively, if the wealthy spend their time in outdoor leisure activities while the poor toil inside factories or offices, then darker skin is considered to be more attractive. In modern industrialized cultures where the average employee works in an office building, paler skin marks being part of the working class. Thus, many people tan to achieve a wealthier, more leisurely appearance. Since skin color is so variable and the ideal color is socially and culturally dependent, perhaps more important is homogeneity of skin color. Having uniform skin color, regardless of shade, indicates health and genetic quality and contributes to the ratings of attractiveness.

Skin Texture

Dr. Fink, Dr. Grammer, and Dr. Thornhill also found that skin texture, another component of complexion, was an important indicator of attractiveness. In fact, skin texture seems to be more influential for ratings of attractiveness than skin color. The men in the study described in the last section rated smooth skin as more healthy and attractive, regardless of race, nationality, or age. Uniformity of the skin can be used as an unconscious indicator of a strong immune system, resistance against disease, youth, and overall genetic fitness.

Skin quality is so important that information about skin quality may be enough to quite accurately interpret attractiveness. When researchers asked individuals to rate a small patch of facial skin for the level of attractiveness, they found that the ratings made with this limited information were correlated with overall face attractiveness. Thus, homogeneity of skin color and quality of skin texture may be more important skin aspects than skin color when seeking an overall attractive and healthy appearance.

From an evolutionary perspective, smooth skin does tend to be strongly correlated with underlying health, so it is an excellent source to gather information about genetic quality. However, skin texture is also an easy trait to modify with cosmetics. In modern societies, there is extensive use of lotions, foundations, and concealers to smooth out skin texture and to create a healthy glow, even in the absence of actual genetic quality, so the validity of skin health as an indicator for genetic health has become less and less valuable.

Tattooing in New Zealand and Papua New Guinea

Homogeneity of skin color and quality of skin texture are two important aspects that significantly contribute to ratings of attractiveness. Interestingly, however, people around the world intentionally disrupt the homogeneity of their skin with tattoos, piercings, and scarification. For example, the Maori tribe of New Zealand and tribes throughout

A woman from the Tufi region of Papua New Guinea has the traditional swirled facial tattoo that marks maturity and attractiveness. Within the Korafe tribe, these symmetrical patterns increase ratings of beauty and signal readiness for marriage. (imageBROKER/Alamy Stock Photo)

Papua New Guinea have independently evolved to have tattooing traditions as part of their cultures.

In the Maori tribe from New Zealand, the men inscribe their faces with permanent, deeply etched, facial tattoos. Although such tattooing procedures should objectively decrease levels of attractiveness because they disrupt skin homogeneity, those with tattoos are consistently rated as more attractive and more desirable. For these men, the visible marks represent strength, number of past kills, and their social status within the group. Men without facial markings were traditionally not allowed to trade, participate in ceremonies, or participate in other tribal activities. These tattoos made them look fierce in battle, frightened their opponents, and were considered highly attractive to women. The tattoos represent accomplishments, mark achievements and identity, and increase their appeal as marriage partners.

The Chambri tribe in Papua New Guinea utilizes a different pattern of markings to adorn their men. This group uses a combination of scarring and pigmentation to create raised tattooed scars that radically change the appearance of their boys and men. In this tattooing process, small incisions made over the torso and buttocks, are rubbed with black ash, clay, and tree oil to create blackened, textured tattoos. Through this painful process, the coloration and texture of the scarring on the torso and buttocks start to take the appearance of crocodile skin. Although the marks permanently disrupt the texture and coloration of the skin, they are seen as indicators of dedication, discipline, and manhood, and thus enhance physical attractiveness. Historically, these tattoos were bloody and painful and men were expected to bear them with stoicism, illustrating their strength and endurance.

In other tribes throughout the Tufi region of Papua New Guinea, it is the women who undergo tattooing procedures. Using charcoal, squid ink, or, more recently, tattoo ink, girls mark their transition into womanhood with permanent facial tattoos. These tattoos consist of symmetrical lines and swirls over the entire face that mark their readiness for marriage. Although such tattoos disrupt the homogeneity of the skin, they increase symmetry, advertise sexual maturity, and are rated as highly striking and attractive by males.

FACE SHAPE AND STRUCTURE

The quality, color, and texture of both skin and hair contribute significantly to levels of attractiveness in humans. Beyond these comprehensive traits, more discrete features also contribute to overall attractiveness. Examination of facial features is a key piece to understanding attractiveness. Although humans commonly use fashion to sculpt, define, or hide the body, an individual's face is commonly on full display, at least in many cultures. Therefore, it makes sense that facial features would play a key role in providing signals of attractiveness to others. Research has found that, in general, biologically sex-specific traits tend to be rated as more attractive by both men and women. Specifically, women who possess highly feminine traits and men who possess highly masculine traits tend to set the standard for what is considered beautiful. For example, for women, a smaller, more pointed jaw is rated as more attractive, while a strong solid jaw line is attractive on a man. High cheekbones, higher eyebrows, and full features also correlate with higher attractiveness ratings for women while more rugged, square features are correlated with higher ratings for males. These features highlight sexually dimorphic traits and provide information about underlying hormone exposure and quality.

Jean-Yves Baudouin and Guy Tiberghien from the Institute of Cognitive Sciences in Lyon, France, present research, which substantiates that various facial characteristics contribute to facial attractiveness. They specifically examined feminine features and identified that large eyes, petite noses, high cheekbones, small chins, high brows, and wide mouths contribute to the ratings of attractiveness for women. In order to understand why these particular features signal attractiveness, these researchers examined how each develops. They found that these features are impacted by hormone levels in puberty and thus provide an indication of underlying physiological functioning. Specifically, females who had a high estrogen-to-testosterone ratio in puberty express these particular characteristics. Thus, facial traits that we see as particularly feminine reveal evidence about hormone levels throughout development.

Face Shape

Dr. Schmid from the University of Nebraska Medical Center describes the ideal face as a ratio of length and width. The ideal ratio for the face is 1.6. That creates an oval-shaped face that is almost twice as long as it is wide. Beyond this golden proportion, there are many other qualities that contribute to individual attractiveness. For example, face shape tends to vary between the sexes. Strong jaw and brow lines, presence of facial hair, and strong features are characteristic of a masculine face. A smaller chin, higher cheekbones, lighter, higher eyebrows, and more delicate features characterize a feminine face. As with other features, much of the research on attractiveness shows that increased levels of sexual dimorphism, or increased masculinity for men and increased femininity for women, increase the ratings of attractiveness for each gender. This research is particularly consistent for ratings of female faces. Feminine attributes in the construction of the female face increase a woman's ratings of attractiveness by others. Furthermore, female faces are rated to be softer, fuller, and the most feminine when the woman is in the most fertile phase of her menstrual cycle.

For men, the research on face shape is more contradictory. Base research does find that participants rate more masculinized faces as more attractive in psychological studies. Other research concludes that more feminized male faces are sometimes rated as more attractive. Still other research proposes that an average male face is most attractive. These differences may be a by-product of methodological differences, a result of cultural differences, or dependent on the phase of the menstrual cycle of the female rater. Victor Johnston and colleagues from the New Mexico State University designed a research study to attempt to parcel apart these different findings. They asked 42 female undergraduate students to rate a continuum of faces for averageness, dominance, attractiveness, health, masculinity, femininity, intelligence, and androgyny. Then the students were asked to rate the photos on 20 different attributes such as overall attractiveness, sexual appeal, trustworthiness, impulsivity, helpfulness, and wealth. Two weeks later, the women were asked to complete the ratings again to see changes in perception over the menstrual cycle.

Dr. Johnston's results showed that women rated more masculine faces (those with broader jaws and larger brow ridges and cheekbones) as more attractive than average male faces. They also rated these faces as healthier. However, as the faces reached the highest levels of masculinity, the attractiveness ratings decreased. The most masculine faces were rated as more dominant and more threatening rather than more attractive. Thus, higher-than-average testosterone, which increases levels of masculinity, is rated as more attractive, but very high levels of testosterone negatively impact the ratings across both physical and behavioral dimensions. Dr. Johnston did find that when women were the most fertile, there was a shift in preference to the most masculinized males, although these faces were also only rated as preferred for short-term relationships, not long-term partnerships. Furthermore, when the female raters were in the less fertile phases of their cycles, they preferred the less masculinized faces.

Although the level of attractiveness changes for male faces depending on fertile phase and relationship goals, these changes are predictable and tied to reproductive phase. When women are the most fertile, they may benefit most from a relationship with a healthy, strong, dominant male. These are characteristics more frequently found in males with higher levels of testosterone who have more characteristically male features. When women are not the most fertile, they benefit more from a caring, nurturing partner, who displays more characteristically feminine traits that result from lower levels of testosterone.

Elongated Skulls in the Congo

The Mangbetu tribe of the Congo in central Africa molds the skulls of their infants to make them long and cylindrical as they grow. This process, known as Lipombo, serves to elongate the back of the head, creating a long cylindrical skull. To create the elongated skull, the Mangbetu take advantage of the fact that the bones of a newborn's skull are not yet fused and still flexible. This flexibility explains why some babies born via vaginal delivery have a cone-shaped head that slowly rounds out after birth. In the Mangbetu, starting within the first month after birth, the infants' skulls are wrapped with tight cords that put pressure on the bones and cause them to shift to form into a long cylindrical shape rather than the normal circular shape. The skulls of the infants are also pressed and wrapped using pieces of giraffe hide or wood to create the desired shape. As the skull develops, the bands are replaced and tightened, sculpting a long, cylindrical skull. This process continues over the first year until the bones of the skull fuse and are less malleable. However, hairstyles are then used to further accentuate the shape of the skull. The result of Lipombo is an elongated skull that symbolizes high social status, high intelligence, and exceptional beauty. The Belgian government, which took control of the areas in the 19th century, outlawed this tradition but its practice has been difficult to control.

Although the procedure of shaping the skull may sound barbaric, research suggests that shaping the skull does not negatively affect the brain. Since the brain grows so quickly after birth and is so malleable, it tends to grow in the space allowed. Thus, the brain may assume an atypical shape as a result of Lipombo, but its processes are not impeded. Research on head elongation has revealed that the practice dates back to as early as 400 BC. Furthermore, the process is shared by many cultures around the world. Skulls demonstrating the process of head elongation have been found among the Huns, among Aboriginal tribes in Australia, and peoples of the Bahamas, as well as Egyptian, Mayan, Incan, and Native American groups. Although the tradition has died

Face Shape and Structure

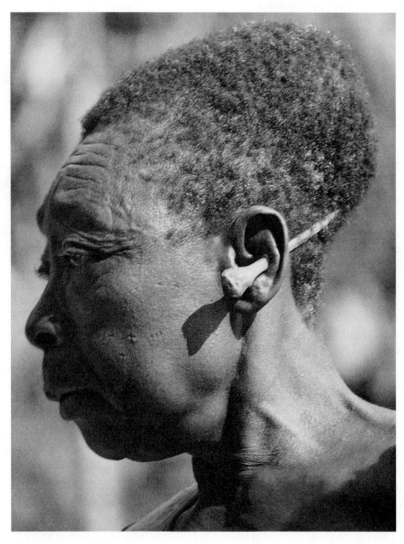

An upper-class Mangbetu woman displays the traditional elongated skull that signals exceptional beauty, status, and intelligence. (Robert Estall photo agency/Alamy Stock Photo)

out in many of these areas, it still continues in the Vanuatu of the South Pacific Ocean. For all groups, the long skulls marked group affiliation, status, majesty, intelligence, and overall beauty.

Cheekbones, Chins, and Foreheads

The structure of the face frames the internal features and contributes to the ratings of attractiveness. For women, in particular, high, prominent cheekbones are positively correlated with the ratings of attractiveness. Men with prominent cheekbones tend to appear more feminine, and are rated as less attractive, at least by women in the fertile phases of the reproductive cycle.

A small, delicate chin is also rated as more attractive for females. A narrow chin is associated with youth, so the size of the chin affects the perception of age. When viewing a female face with a small chin, there is a rise in brain activity for the male viewer, indicating his increased attention and interest. As a photograph of a female is manipulated to make her chin appear wider, her ratings of attractiveness proportionally decrease. For males, a larger, more structured chin is considered to be more dominant and attractive and a small chin is associated with immaturity and a baby face. A square, solid jaw indicates strength and masculinity and is rated as particularly attractive, especially during a woman's most fertile phase.

Although these individual structural features were found to contribute to overall attractiveness, their placement and proportion matter as well. Furthermore, the preferred placements and proportions varied depending on several qualities of the viewer. For example, the height of the rater predictably impacted their preferred proportions. Since familiarity influences the attractiveness ratings, the height of the viewer was found to influence ideal ratios of the viewed individuals. Taller individuals tended to select those photographs with larger foreheads and smaller chins to be more attractive. When viewing people from above, their foreheads would appear larger and their chin smaller, simply due to perspective, making these proportions more typical for that viewer. Shorter individuals preferred larger chins and smaller foreheads, representing what they typically see when looking at others from their shorter vantage point. In fact, in a research paradigm, the ideal feature placement and size varied predictably by the height of the viewer. Mothers, in particular, preferred larger foreheads and smaller chins, likely as a by-product of their attention on small children who are lower in their fields of vision. This preference was shared by adolescents who spent time engaged with children, but was not shared by college students who spent more time with people at a more level vantage point.

Cheekbones, chins, and foreheads create the overall structure of the face and contain distinctive differences between the genders. These traits are differentially created by underlying hormone exposure and are correlated with genetic health. For example, traits such as strong chins and smaller eyes, that are typically rated to be more masculine and more attractive during a woman's most fertile phase, are correlated with better male health and increased reproductive success. Thus, men who display such traits are likely to garner more attention and have more opportunity to reproduce.

EYES

Eye Color

Eye color is a trait that is highly researched with respect to attractiveness, with varying success. Bruno Laeng and colleagues from the University of Tromso in Norway showed participants photos of males and females with varying eye colors. Female participants and brown-eyed males showed no stable preference in their ratings of attractiveness of blue- or brown-eyed photographs. However, blue-eyed males showed a clear and stable preference for blue-eyed females. When examining these results, Dr. Laeng hypothesized that, since blue eyes are a homozygous recessive trait or are only expressed when the individual receives the genes for blue eyes from both parents, they provide a reliable clue of the underlying genotype. Blue eyes typically only occur when the individual has two alleles coding for blue eyes, so this expressed eye color is a reliable clue to the genetic makeup of the individual. This makes blue eyes unique in comparison to darker eye colors. Dr. Laeng thus suspected that blue-eyed males would have an innate tendency to prefer blue-eyed females because the resulting children would give clear indicators of paternity. A blue-eyed male and a blue-eyed female should have blue-eyed children because the gene for blue eyes is the only form of the gene that they each have to pass on to their offspring. Thus, a brown-eyed child would quickly reveal infidelity to a blue-eyed male.

Although ratings of attractiveness revealed that blue-eyed males rated blue-eyed females as more attractive, Dr. Laeng was interested in whether or not this expressed preference carried over into actual dating behavior. In a follow-up study examining actual partnerships, Dr. Laeng and colleagues found a similar pattern in actual behavior. Blue-eyed males tended to be in relationships with blue-eyed females. From an evolutionary perspective, this would provide increased confidence of paternity for those males because their genotype would show in their offspring's phenotype. This certainly does not mean that blue-eyed males are consciously choosing females with a thought to future confidence of paternity, but it could be a by-product of successful reproductive efforts in our evolutionary history. Blue-eyed males from our evolutionary past, who had a natural preference for blue-eyed females, were more likely to successfully pass on their genes and invested more of their resources in children who actually carried their genetic material, thus passing this preference on to future generations. Blue-eyed females do not have the same evolutionary pressures as males because females can be confident that the child they carry is their own offspring. Women are not as reliant on physical traits to prove genetic relatedness and can, thus, mate with anyone of their choosing.

Eye Placement and Gaze

Although research seeking a universal preference for eye color has not yielded consistent results, other aspects of the eyes have been more revealing. Eye placement, including height on the face and distance between the eyes, and eye symmetry predictably affect the ratings of attractiveness. Eye gaze, similarly, serves as a marker of attraction and influences the ratings of attractiveness.

Dr. Jean-Yves Baudouin and Dr. Guy Tiberghien from the Institute of Cognitive Science examined the contribution of these individual features on measures of attractiveness for women's faces. They presented male subjects with images of pairs of women's faces and asked the men to select which face was more attractive. While they found that asymmetrical faces were rated as less attractive, asymmetry of the eye region was one of the key areas that impacted the attractiveness ratings. Specifically, eyes that were asymmetrical in height on the face or not the same distance from the center of the face negatively impacted the attractiveness ratings. Asymmetry of the eyes may provide clues of poor underlying genetic health or of poor stability in the developmental environment.

Dr. Chihiro Saegusa and Dr. Katsumi Watanabe from the University of Tokyo also illustrate the importance of the eyes in the ratings of facial attractiveness. In their research, they asked men and women to rate female faces and facial features for the level of attractiveness. Unlike ratings for the nose and mouth, they found that ratings of the eyes were consistent regardless of how long the participants were allowed to view the photographs. Dr. Saegusa and Dr. Watanabe concluded that the eyes more significantly contribute to the perception of attractiveness than any other facial feature. Furthermore, when gaze was averted, attractiveness ratings were lower, indicating that eye contact increases attractiveness ratings. Due to the high communicative value of eye gaze, the connection made through eye contact increased the overall ratings of attractiveness for the whole face.

Eye Size

In addition to the placement and color, eyes vary in size. At birth, eyes are essentially fully grown, making the eyes proportionally larger in the face. Thus, babies' eyes appear large and as the face grows, the eyes become more proportional to the face. The ratio of the eye to the overall face can therefore be used as an indicator of age. Specifically, larger eyes communicate youth and immaturity, and adults with naturally larger eyes tend to look younger.

> ### Double Eyelid Surgery in China
>
> Around the world, the features of the eyelid vary from a single lid, without a naturally occurring crease, known as a single eyelid, to a lid where there is a crease in the small flap of skin over the eye, known as the double eyelid. In a medical research study looking at the attractiveness of different types of eyelids, the double eyelid was endorsed as the ideal of beauty and the single eyelid was considered to be the least attractive. Dr. Hwang and Dr. Spiegel altered photographs of women to create low, medium, and high upper eyelid creases and both Asian and non-Asian raters considered the medium level of eyelid crease to be the most attractive and an absent eyelid crease was the least attractive.
>
> The typical type of eyelid varies between cultures, but for those of East Asian descent, only about 50 percent of people have a naturally occurring upper eyelid crease while the others have only a single lid. Despite how common the single eyelid is in China, the double eyelid is coveted as a more beautiful trait, particularly for women. In fact, in Asian cultures, a double eyelid is identified as a hallmark of beauty. In response to the common lack of an eyelid crease, an increasing number of Chinese women are seeking surgery to create the double eyelid, a procedure known as an Asian blepharoplasty. In this $3,000 surgery, a surgeon cuts, folds, and stitches the eyelid skin to create a crease. This process pulls the eyelid higher when the eyes are open, exposing a larger portion of the eye. This makes the eye appear larger and rounder, and gives the patient an appearance more typical of European or American women. This surgical procedure to create a double eyelid has grown to be the most common cosmetic surgical procedure in South Korea, Taiwan, Japan, and East Asia and is commonly given as a graduation gift from parents to young women as they prepare to enter the professional world.
>
> Through this invasive surgical procedure, patients report the hope of appearing more energetic, confident, and attractive. Most of these Asian women do not report an interest in looking more like Europeans or Americans but want to look more like other Asians who were born with a crease. Following the surgery, patients report feeling more attractive, having more confidence, having better dating opportunities, and having greater job prospects.

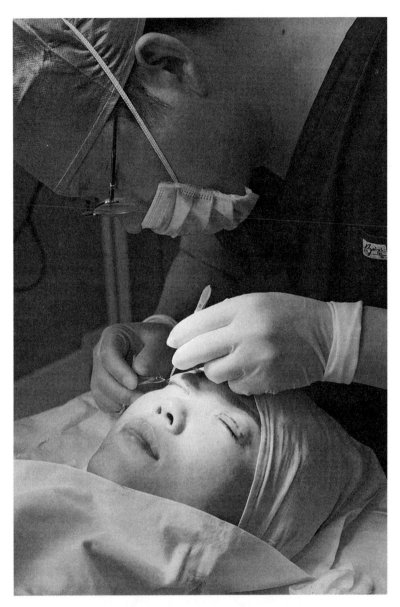

A young professional nurse undergoes double eyelid surgery in a quest to increase physical beauty, self-confidence, and future job prospects. Double eyelid surgery is the most common cosmetic surgical procedure throughout Asia and creates the appearance of a double eyelid that is more common among non-Asian populations. (Mike Goldwater/Alamy Stock Photo)

Since youth is correlated with feminine beauty, female faces with larger eyes tend to be rated as more attractive.

Dr. Caroline Keating from Colgate University found that more mature features (e.g., small eyes, prominent jaw, increased facial hair, thick eyebrows, thin lips) made male faces more attractive and more dominant. The same features led to lower ratings of attractiveness for female faces. Female faces were rated as more attractive when they displayed immature facial cues (large eyes, small chin, thin eyebrows, plump lips). Dr. Keating found that eye size was the biggest predictor of both dominance and attractiveness ratings. Eyes that made females look younger and less dominant led to higher ratings of attractiveness. Large eyes made faces look more submissive, thus women with larger eyes were consistently rated as more attractive. Eyes that made males look more dominant led to higher ratings of attractiveness. For men, smaller eyes made the males look more dominant; thus men with smaller eyes were rated as more attractive.

Iris, Pupil, Sclera, and Limbal Ring

Attractiveness research focused on the eyes is significant because individuals spend 60 percent of the time looking at the eyes when interacting with others, particularly when attempting to remember or recognize a face. Humans tend to direct attention to the eyes more than any other feature. Research has demonstrated that besides size, color, placement, and gaze, there are other aspects of the eyes that contribute to attractiveness. These include the size of the iris, the diameter of the pupil, the color of the sclera, and the prominence of the limbal ring.

The iris is the colored part of the eye. Although the color may not be conclusively predictive of the level of attractiveness, its width does seem to play a role. Over the life span, the colored part of the eye decreases in proportion to the overall amount of the eye displayed. Thus, younger individuals appear to have larger irises in comparison to the overall eye. This makes the colored part of the eye appear larger during youth and this larger proportion of color is attractive to males. With age, the proportion of the visible eye that is taken up by the iris decreases. This decrease in size may make the individual appear more mature and more dominant, leading to decreased ratings of attractiveness for women but increased ratings for men.

Pupil diameter is also correlated with the ratings of attractiveness. The pupils tend to increase in size when one is looking at something interesting or when one is sexually aroused. Larger pupils may make the eyes look bigger and may make people appear more interested in whatever it is that has captured their gaze. Large pupils tend to be rated as more attractive by males, likely because they are viewed as more interested and potentially more aroused. Women's rating of pupil attractiveness is less consistent. Women do not rate larger pupils as more attractive except for when they are in the most fertile phase of their menstrual cycles. This increase in attraction may be due to an increased interest in sexual activity and more attention devoted to securing a male partner.

The sclera, or the white part of the eye, also contributes to the ratings of attractiveness. The sclera naturally changes from a bluish tone at birth to a whitish hue in adulthood and then slowly yellows with age. An individual with an unblemished, bluish or white sclera is rated as more attractive than when the sclera is manipulated to have yellowing or reddish hues. The tone of the sclera tends to become more yellowed with age and sickness. Furthermore, sadness or exhaustion can cause it to have a reddish appearance. Overall, discoloration tends to be correlated with disease or declining health and is rated as less attractive.

A final feature of the eye that contributes to the ratings of attractiveness is the limbal ring. The limbal ring is the darker ring of color along the outer edge of the iris (the colored part of the eye). Individuals with a darker limbal ring appear younger and are rated as more attractive. The darker color may also create a contrast with the white of the eye to making the colored part of the eye appear larger and more striking and making the sclera appear lighter and healthier by contrast.

NOSE

The ideal nose varies between the genders, but symmetry and averageness are key aspects contributing to the overall attractiveness for both men and women. Even without research, one can safely assume that the nose is important in overall attractiveness due to how common it is as a target for cosmetic surgery. As the central feature of the face, its placement, size, and structure can significantly alter the overall attractiveness of the face. Research on facial attractiveness has revealed that the nose is more important than many other features, including the mouth, for determining facial beauty.

For women, research has revealed that there is no one ideal nose. There are lots of different shapes and sizes that can contribute to increased levels of attractiveness. Female noses can be perky, regal, button, pert, or delicate. How well the nose fits with the shape and other features of the face determines whether it contributes or detracts from overall attractiveness. However, in general, noses that are slightly smaller and narrower than average are considered the most attractive for female faces.

For males, however, there does seem to be an ideal nose that contributes to the overall ratings of attractiveness. For men, a shaped, structured nose that lends a sense of confidence and dominance tends to increase the ratings of attractiveness. Ideal male noses tend to be bigger, less delicate, and more strongly defined. The ideal male nose protrudes directly out from the face so it does not look upturned nor droopy. It should be broad and not pinched off at the tip and should err on the side of convex rather than concave. A concave, pinched, upturned nose creates a feminized look that detracts from the ratings of male attractiveness.

Since the nose continues to grow throughout the life span, smaller noses make the face appear younger and larger noses are an indicator of increased age. This may be one reason that smaller noses are more attractive for women. The smaller nose gives the impression of youth, an attractive female trait. For men, larger noses may indicate increased maturity, an attractive male trait. Due to its impact on facial attractiveness, rhinoplasty is a common cosmetic surgical procedure. In 2015, almost one million individuals from 10 different countries had their noses reshaped. Rhinoplasty makes up almost 10 percent of all cosmetic surgical procedures. In the United States, men make up only about 10 percent of the surgeries, but on an international scale, this number rises to 30 percent.

EARS

Research has demonstrated that ears do not enhance or detract from overall attractiveness to the same extent that the internal facial features do. Only extreme variations in ear size, placement, or symmetry have an influence on attractiveness ratings. Properly formed, normally placed, averaged-sized ears tend to be the most attractive. Ears that deviate from the norm can detract from attractiveness, but typically, only ears that are placed extremely low, extremely high, or stick out excessively actually have an influence. It seems that ears, by themselves, are more functional than aesthetic. Individuals in many countries adorn the ears with jewelry or gauges to artistically frame the face, so ear adornments may add to decoration and attractiveness even if the ears themselves go unnoticed.

Stretched Earlobes in Kenya

The Maasai ethnic group is a semi-nomadic people who live in southern Kenya and northern Tanzania. Among other things, they are known for their bead working, body painting, body ornamentation, and body modification. In particular, stretched earlobes are considered the most beautiful among the Maasai for both men and women. As young men and women take on the responsibilities of adulthood, their earlobes are pierced and then slowly elongated. Natural materials, such as increasingly large thorns, stones, bundles of twigs, or elephant tusks, can be used as gauges to slowly stretch the earlobes over the life span. Alternatively,

A senior Maasai man from Entasekera, Kenya, exhibiting the traditional stretched earlobes that indicate status, maturity, and physical beauty. His ears are also decorated with strands of beads to complete the adorned look. Longer earlobes garner greater respect within the group. (Sjors737/Dreamstime)

weights can be hung from the piercing to cause the earlobes to elongate. Since the stretching process takes time, longer lobes symbolize age, maturity, and wisdom and garner greater respect within the group.

For the Maasai men, longer earlobes are used to indicate warrior status. Although Maasai women do rate the stretched lobes as more attractive, the earlobes are also used to signal success in hunting and success in fighting, and serve to be intimidating to outsiders. These longer lobes also correlate to higher status within the group. The higher-status males are rated as more attractive to the females and more sought after as marriage partners. For the women of the Maasai, extravagant beaded ear ornaments signify marriage status and beauty. Unmarried females wear beaded earrings that are specific to their unmarried and available status.

For both men and women of the Maasai, stretched earlobes signal attractiveness, maturity, status, and reproductive readiness. In the stretched lobe, individuals sometimes wear plugs, but more often, once the earlobe has been stretched, the hole is decorated with hanging strands of beads. During ceremonies, these beads swing in time with music or dancing, to draw attention and to increase ratings of attractiveness. In addition to the stretched lobes, the men and women also hang strands of beads from piercings in the tops of their ears to finish out their adorned look.

In the modern Maasai societies, more women than men continue the tradition. Today, travelers can visit the Maasai to experience their culture and to see the body modification practices. Although exposure to the outside world through visitation is slowly impacting the culture of the Maasai, the tourism also allows the Maasai to educate visitors about the importance of land and animal conservation and provides funds to help maintain their way of life within a developing world.

LIPS

Due to their communicative value, the lips and eyes are the focal points during almost all face-to-face conversations. Most of the time spent looking at another's face is split between the eyes and the lips. Lips deliver speech, communicate emotions, and are a key piece in kissing. Given this wide array of activity and high prominence in day-to-day life, it is not surprising that research has found that lips are highly influential for ratings of sexiness and attractiveness. Plump, full lips can be particularly pleasing, and a rich contrast of color between the skin

Stretched Lips in Ethiopia

The Mursi women of Southern Ethiopia insert clay plates into their lower lips to permanently elongate their lip. As a girl enters adulthood, a half-inch incision is made in her lower lip and a wooden peg is inserted to hold it open as it heals. Once the incision heals, a larger peg or clay plate is inserted to slowly increase the size of the opening. The resulting stretched lip is considered a sign of status, maturity, and beauty. The lip plate is used to indicate sexual maturity, strength, identity, and attractiveness. Adult women use the lip plates to accessorize their appearances and wear different plates for different occasions.

While completing field research as a doctoral student, Shauna LaTosky explored the significance of the lip plate for Mursi women. She found that the lip plate is tied to social status. Unmarried and newly married women are the most likely to wear the lip plates, and older married women with children wear them only for special events. LaTosky found that the significance of the lip plate is to signify fertility and eligibility for marriage. The lip plates garner respect from men, and those with lip plates are rated as more attractive and of higher status, and are more sought after as marriage partners. LaTosky likened the lip plates to the European and American fashion of stilettos. The Mursi women wear the lip plates to increase attractiveness and are more likely to wear them for formal events. They also tend to wear it prior to events so that it will be comfortable for the event, similar to how women need to wear stilettos around the house to make them more comfortable before an evening out. Furthermore, Mursi children pretend and wear bent branches in their mouths with leaves to simulate a lip plate during play.

Although older Mursi women do not tend to wear lip plates as frequently, they may wear them for special occasions such as for ritual events like weddings, competitions, and dances. They also wear them to pose for photographs with tourists to supplement income. Modern Mursi women who no longer choose to wear a lip plate or do not cut their lips at all risk being perceived as lazy, are less respected by men, and may not achieve the bride wealth that a woman with a lip plate would receive. Thus, family pressure, social disapproval, and public perception all help maintain the tradition.

A young Mursi woman wears the traditional clay lip plate in her lower lip. The stretched lip signals maturity, strength, and great beauty. Women who conform to this tradition are more respected and more sought after as marriage partners. (Uros Ravbar/iStockphoto.com)

and the lips draws attention, especially on a woman's face. Lips play such a significant role in the overall facial attractiveness that they are common targets for cosmetic surgery.

To look at contrast and color of the lips, Ian Stephen, from the University of Bristol, and Angela McKeegan, from the University of Ulster, present research that illustrates the attractiveness value of lips for men and women. In their research, they report on research findings that demonstrate that increased contrast between facial features, even in black-and-white photos, increases the appearance of femininity but decreases ratings of masculinity.

To see if color matters, Dr. Stephen and Dr. McKeegan instructed college students to adjust the color of lips on male and female photos to enhance the individual's attractiveness. Students were specifically instructed to either make the faces as attractive as possible or to make the faces as masculine or feminine as possible. For faces of both genders, participants increased the redness in the lips when instructed to maximize attractiveness. When attempting to maximize femininity or masculinity, however, participants increased lip redness for female faces and decreased redness for male faces. This natural preference for redder lips in females likely contributes to the cross-cultural, cross-temporal use of lip reddening cosmetics.

As with many other physical features, lips are a sexual dimorphic trait. Women, with increased levels of estrogen, tend to have larger, fuller lips than men. Thus, larger, fuller, redder lips may communicate underlying health and vitality and attract a male's attention quickly and consistently due to their prominent display. For males, redder, fuller lips are not needed to communicate underlying hormone levels, so males tend not to accentuate that feature when looking for a mate.

It is no wonder that women in many cultures make significant attempts to plump and color the lips and use jewelry such as lip rings or lip plates to further draw attention to this part of the body. When examining photos of females, red lips are consistently rated as the most alluring, and are correlated with ratings of good health and youth. In the United States alone, there were more than a million lip augmentation procedures completed within the last year. Lips can be injected with fat or gels to enhance their size and shape, or chemicals can be used to temporarily irritate, redden, and swell the lips to provide a plumper appearance. These procedures have been sought almost exclusively by women in attempts to increase the sexual appeal of their faces.

TEETH

Oral hygiene is a remarkable indicator of overall health, and the quality of one's teeth directly affects one's ratings of attractiveness. Tooth health, particularly in eras or regions of the world without modern medical care and hygiene tools such as toothbrushes and toothpastes, directly reflects diet, stress levels, and genetic quality. Left untreated, dental decay will contribute to tooth discoloration, tooth loss, bad breath, and poor health. Such decay will eventually lead to infection, and potentially death.

In industrialized cultures, many individuals have some access to dental care, so tooth health becomes a less reliable indicator of underlying genetic quality. Even someone with very poor dental health may have access to treatments that mask the natural quality of his or her teeth. Left untreated, however, poor dental health is impactful for ratings of attractiveness just as it has been throughout history. And, with the modern medical climate, poor dental health is particularly detrimental to the ratings of attractiveness. Not only does poor dental health indicate poor genetic quality but it also indicates poor grooming habits, laziness, or lack of access to medical care. Thus, even with quite widespread dental care, tooth appearance still has an exceptional impact on the ratings of attractiveness.

Smile attractiveness is highly correlated with facial attractiveness. When communicating, individuals spend a significant amount of time looking at the mouths of their partners. A nice smile is correlated with higher ratings of intelligence, social ability, and attractiveness. The attractiveness of the smile is dependent on its size, shape, and placement; color of the teeth; and the shape and placement of the lips. The lips frame the smile and more teeth tend to show in a youthful smile. The lips tend to sag with age, and provide fewer glimpses of the teeth beneath.

In a study by Pieter Van der Geld, Paul Oosterveld, Guus Van Heck, and Anne Marie Kuijpers-Jagtman of Radboud University Nijmegen Medical Centre, results showed that when individuals viewed and rated spontaneous smiles (smiles that were genuine and not posed for a camera), tooth size, visibility, and upper lip position were most predictive of ratings of attractiveness. A lip that extended far enough up to expose the teeth and a small part of the gums created the most attractive smile. Too much gum exposure, however, decreased the attractiveness ratings. Tooth color emerged as a factor that was more associated with one's satisfaction with his/her own smile rather than with attractiveness, as will be discussed in the next section.

This group of researchers found significant correlations between ratings of smile attractiveness and personality traits. Attractive smiles were positively correlated with self-esteem and negatively correlated with traits of neuroticism. Visibility of teeth during a spontaneous smile and the position of teeth were also positively correlated with dominance and assertiveness. It is possible that a certain level of self-esteem, self-assurance, and dominance is needed to allow oneself to engage in a full-toothed smile. Holding back on a full smile may signal stress, self-consciousness, and submissiveness.

Therefore, if confidence and dominance are needed in a situation, using a full smile could provide a positive first impression.

Tooth Color

Alexis Grosofsky and colleagues from Beloit College took a closer look at tooth color to see if color has an effect on attractiveness. The American culture, in particular, is seemingly obsessed with tooth-whitening agents, and the media is saturated with whitening toothpastes, strips, and dental advertisements. Dr. Grosofsky and colleagues found that 40 percent of the American population are unhappy with their teeth and that two of the five most common dental procedures are whitening and laminates. By 2000, about $1.5 billion was spent each year to alter tooth color.

In contrast with this trend, research on the impact of tooth color repeatedly demonstrates that tooth whiteness is not actually correlated with health and attractiveness ratings. Given the increasing focus on whiter teeth in our culture, Dr. Grosofsky believed that whiter teeth would be rated as more attractive in a psychological study. However, she and her colleagues found no significant difference in attractiveness ratings based on tooth color. The question then becomes, why are people so obsessed with whitening their teeth? The trend for whiter teeth is likely a by-product of advertising and social influences. Media tells us that whiter teeth are more attractive. Thus, individuals who have or obtain whiter teeth may also feel more attractive, which would have a positive impact on self-esteem, which increases confidence levels. Confidence has been shown to influence how and how frequently one interacts with others as well as how one is perceived by others and can definitely have an impact on attractiveness ratings, totally unrelated to actual assessment of tooth color. Furthermore, teeth tend to darken with age, so whiter teeth are associated with youth, and thus it follows that whitening the teeth may make an individual appear younger. This may also have an indirect impact on the attractiveness ratings.

Colin Hendrie, from the University of Leeds Institute of Psychological Science, and Gayle Brewer, from the University of Central Lancashire, also looked at the role of tooth color in the ratings of attractiveness. They recognize that over $1 billion per year is spent on cosmetic dental procedures and that such procedures are sought out cross-culturally. In their study, they altered the appearance of teeth in photographs to see the impact on the ratings of attractiveness. They recognize that teeth exhibit both aspects of biology and environment, including disease and trauma. They found that abnormal spacing and extreme yellowing of the teeth had an impact on the ratings of attractiveness, particularly for photos of women. However, normal discolorations were rated similarly to whitened teeth. Only abnormally yellowed teeth impacted the ratings, regardless of the sex of the rater. Since teeth tend to darken with age, extremely yellowed teeth may be used as a sign of age and decreased reproductive viability. This would explain why yellowed teeth have a higher impact on ratings of female photographs than male photographs. Women also rely on their teeth more in social displays by smiling more and wearing bright lipstick to enhance the prominence of the display. Such smiling displays become associated with female attractiveness, and thus there is more of an impact when the display is disrupted or abnormal.

So, for those preoccupied and self-conscious about the color of their teeth, it may be relieving to note that the procedures to whiten the teeth do not have an impact on overall attractiveness. Unless one's teeth are abnormally darkened or yellow, such as by smoking or disease, there is no benefit to enduring procedures to whiten the teeth. The benefits of teeth whitening may only lie in self-esteem, self-confidence, and self-perception, rather than in the perceptions of others.

HANDS

Krzysztof Kosciusko from the Institute of Anthropology in Poland contributes to the research on attractiveness by examining the attractiveness value of hands. Although hands are not as prominent and visible as the face and overall body shape, hand appearance is found to be important in overall ratings of attractiveness. Given these findings, it is unsurprising that women tend to adorn their hands with rings, nail polish, lotions, and bracelets. For women, smaller hands with long, delicate fingers tend to be rated as more attractive. Also, the smoothness of the skin can increase or decrease hand attractiveness.

For men, larger hands tend to be rated as more attractive. Interestingly, finger length is directly tied to the level of testosterone exposure in the womb. High testosterone exposure results in longer ring fingers, in comparison to the index finger. High testosterone is correlated with higher sperm count, greater fertility, better heart health, and greater facial symmetry. Interestingly, and maybe due to opportunity, these individuals are also more likely to be promiscuous and engage in more short-term relationships. Men with index and ring fingers of more similar lengths are more likely to be faithful and engage in long-term relationships.

Dr. Kościński was interested in investigating whether hand shape and attractiveness are correlated with attractiveness in other parts of the body. Are the hands an accurate cue to overall facial and body attractiveness? He hypothesized that hand attractiveness would be influenced by hand shape, size, skin quality, suppleness of skin, nail health, and overall fattiness, all cues to levels of femininity or masculinity. Furthermore, he thought that hand attractiveness would be correlated with attractiveness of other body parts. For his study, he photographed the faces and hands of 188 men and women for analysis. He found that many factors contributed to hand attractiveness and those same factors contributed to facial attractiveness. For women, femininity, low fattiness, skin health, and grooming contributed to the ratings of attractiveness. Attractiveness ratings of hands also correlated with facial attractiveness ratings in men. For both genders, averageness was a reliable cue to stability and good genes, and thus, the hands revealed such underlying genetic health. However, environmental factors were also found to contribute. Diet, grooming, and exercise were all correlated with hand and nail health and impacted attractiveness ratings.

BODY SHAPE AND PROPORTIONS

Body Mass Index

Body mass index (BMI) compares body mass to overall height and categorizes an individual as underweight, normal weight, overweight, or obese. What is considered normal weight varies between males and females, with women tending to naturally have more body fat than men. When rating female attractiveness, females express a personal preference for a lower BMI even though males rate females with average BMIs as most attractive. When rating male figures, men and women tend to rate males with higher BMIs as more attractive, at least to a point, making BMI preferences markedly different between the sexes. The differences in preference for BMI between the sexes are likely the result of both evolutionary and cultural preferences. There is a peak BMI for reproductive success, which does not necessarily match the ideal BMI from a cultural perspective. This dichotomy affects body image and psychological health for men and women, particularly in body-conscious societies such as the United States.

Although women rate female bodies with lower BMIs as more attractive and think that they are more successful and more sought after, fat reserves are a necessity for reproductive success for females. Due to reproductive demands, women naturally carry more fat, particularly around the hips and thighs, than normal weight men. Gestation and lactation place demands on the female body, and during those times, the female body draws support from fat stores. Despite the need for fat stores for healthy female reproductive success, many women in industrialized societies strive to maintain a low body weight and most women report feeling overweight and express an interest in

Body Shape and Proportions 205

losing weight, even if they have a BMI in the healthy range. The likely culprit for this phenomenon is media portrayals of the female body. Models tend to have a BMI in the underweight category and are also Photoshopped and retouched to remove any signs of fat or imperfections. Media has the effect of instructing individuals of what is "normal" even if what is being portrayed is in the abnormal range. Thus, women see these underweight models as the normal standard and use them to form their own self-conceptions. Some companies are attempting to reverse this trend by using more typical women in their advertisements and by reducing the amount of Photoshop used to alter the photos, but the extremely thin, young, and beautiful ideal still dominates the advertisement industry.

Although it may be surprising to most Americans, there is a marked preference for heavier women in nonindustrialized societies. This is especially seen in countries where there are risks of food shortages. In these societies, fat reserves are necessary to sustain a child after birth, and heavier women are considered to be more attractive, more fertile, and more sought after for a marriage partner. This is directly in contradiction to the American ideal that places such high emphasis on low BMI and low body weight for females that models and actresses actually tend to be underweight on the BMI scale.

To examine if this trend for underweight ideals is found in other nations, a study including British, Kenyan, and Ugandan participants sought to determine male preferences around the world. Men were asked to rate 12 line drawing figures of women for attractiveness. These figures ranged from underweight to obese. Contrary to the skewed perceptions of American women, none of the males identified the underweight individuals as the most attractive. All groups rated the normal-weight figures significantly more attractive than either the underweight or obese figures. Even more markedly, the Kenyan and Ugandan men preferred heavier figures as compared to the British men. The heavier figures were rated as more physically attractive, healthy, fertile, young, and preferable as spouses. These cultures have little experience with obesity, so it may be the case that knowing the detrimental effects of too much weight did not skew their preferences. Interestingly, however, even at American universities, extreme thinness was not rated as more attractive by males (contrary to what women tend to believe). Since women view female models in magazines, TV, movies, and advertisements, they may be selectively influenced by the trend of low body weight and see it as an ideal, even if the men around them do not.

This is not to say that BMI does not influence the attractiveness ratings. BMI has been found to account for up to 80 percent of the variance of female attractiveness ratings. In a university setting, women and men rated the attractiveness of photos of real women. BMI had a predominant influence on the ratings from both sexes. The women rated as most attractive were those with a BMI of around 19. A BMI of 19 is predictive of women of average body weight. These women were not underweight (BMI < 18), overweight (25–30), nor obese (>30). Neither gender preferred the lightest or heaviest groups.

Large Bodies in Mauritania

Throughout most cultures, an average body mass index of 19 and a low waist-to-hip ratio are rated to be the most attractive for women. However, in the Mauritanian population in Northwestern Africa, overweight women are considered the most beautiful and the most desirable. Girls as young as age seven may be sent to camps where they are forced to eat up to 16,000 calories a day. At such camps, girls are forced to overeat and to drink gallons of camel's milk. By age 10, most are unable to run and have bodies draped in loops of fat. Once they are obese, they are rated as more attractive and more appropriate for marriage partners.

Mauritanian women fetch water from a well in the village of Barkeol, Mauritania. In this area of northwestern Africa, heavier women are considered to be the most beautiful. (Georges Gobet/AFP/Getty Images)

Due to such camps, over 50% of women in this region are obese and many experience high risk of diabetes and heart disease. Because of these health consequences, the federal government has attempted to end the force-feeding tradition, and for a while, the rates dropped so that only one in ten women under the age of 19 had experienced such practices. This was down from 30% of women aged 40 and older. Recently, however, due to fallout with the federal government, there has been a return to the tradition of fattening, and since 2008, force-feeding is again gaining in popularity.

Female flesh is reportedly comforting and erotic to men and stretch marks are the ultimate sign of beauty. Girls and women report using drugs to gain weight and use hormones more traditionally used to fatten camels and chickens. Thin girls are teased and thin women are rated as less desirable as romantic partners. Women report desiring to gain weight to feel more confident and to be more attractive. Although they may have rebelled against their mothers' pressures to gain weight when they were young, many women report seeking out means to gain the weight themselves later in life.

Although a desire to be obese is the antithesis of the U.S. ideal, it makes sense in light of the environmental pressures in the nomadic world. Survival in the desert is difficult and extra weight allows for survival even during times of drought and famine. In this culture, obesity is a sign of prosperity and wealth. Excess body weight demonstrates that the family has access to surplus food even in a country frequently plagued by famine. Thus, to the Mauritanian men, a fat wife symbolizes wealth and success.

Many women in the United States and in other industrialized countries have a marked preoccupation with their weight. For those women, knowledge of research findings might help increase self-esteem and ease body consciousness. First, although BMI may play a role in attractiveness levels, waist-to-hip ratio (WHR) has been found to be a more important than overall BMI for women when predicting the level of attractiveness. Overall body size does not seem to have the overarching impact that most women believe. Body shape seems to be able to more accurately predict attractiveness ratings. Second, although the Western ideal is to be slim, men repeatedly endorse a preference for a larger woman than most women predict. When asked to pick which line drawing of the female figure a man would find most attractive, women choose a significantly slimmer figure than do men. Third, body size preference varies by socioeconomic climate. In cultures where there is plenty of food, slimmer women are preferred when compared to cultures where there are limited food sources. Weight is used as a symbol of status and wealth. When there is food scarcity, a larger woman signals the status of having plenty to eat. In times of plenty, women with a more average body size are preferred.

Although BMI does not seem to have such an impact on male self-esteem as it does on females, it still has an impact on a male's ratings of attractiveness. For men, increased BMI, at least to an extent, correlated with increased levels of fertility. Increased BMI in men is correlated with strength, dominance, and status. Men with larger BMIs tend to have an increased ability to protect their families, to be formidable opponents, and to fight and dominate other men more effectively. Furthermore, the androgens that produce larger body size for males also have a positive impact on levels of aggressiveness, speed, endurance, strength, and muscle mass. Men with higher BMIs also tend to have more exceptional spatial abilities and enhanced cardiovascular efficiency. Thus, men with larger body mass tend to be more successful, have more opportunities, are healthier, and are more attractive to women. Certainly, men who are too large or have too much body fat in ratio to muscle mass see contradictory results. Obese males, for example, would have more difficulty moving, hunting, or fighting, and may have decreased cardiovascular health due to the demands such weight puts on their bodies, so there is a balance point between ideal size for reproduction and ideal size for effective hunting.

Waist-to-Hip Ratio

Although BMI has generally been used as a marker for health and attractiveness, WHR actually seems to carry more cross-cultural consistency for predicting ratings of attractiveness, at least for women. Although ideal BMI has decreased in wealthier and more body-conscious societies in line with the emergence of media portrayals of women, WHR has remained constant as an indicator of attractiveness. Cross-culturally, men prefer an overall WHR between 0.67 and 0.80. Although there is some cross-cultural variation within this range, most men rate 0.70 as ideal. A female figure with a WHR of 0.70 would have the classic hourglass shape that is associated with a feminine body shape. Not only are women with WHRs within the 0.67–0.80 range rated as more attractive cross-culturally, but having this body type also tends to have other benefits. Women with WHRs within this range tend to be biologically healthier than women with higher or lower WHR. A WHR within this ideal is correlated with increased ability to successfully reproduce and with overall health. These women tend to show increased resistance to disease, are less likely to develop cancer or diabetes; have decreased risk of developing hypertension or diabetes; and are less likely to have heart attacks, strokes, or gall bladder problems. Thus, although men may not be conscious of these correlations, those males who choose females within this range are more likely to have healthy and numerous children, thus passing on this preference.

Biologically, WHR develops during puberty as a result of the activating effects of hormones. These hormones impact the development of a woman's thighs, hips, and buttocks, and create visible signals of a woman's underlying physiological functioning and reproductive viability. During puberty, for girls, the pelvis widens, expanding the hips, and creating an even lower WHR. This low WHR is correlated with reproductive potential, and the hormones also contribute to the increased health effects described above, independent of overall level of body fat or BMI. The ideal WHR of 0.70 directly impacts the ratings of attractiveness and sexiness, but its significance goes far beyond mere beauty. WHR really communicates youth, level of health, and reproductive potential.

During pregnancy, WHR increases. A pregnant woman is more tubular and has less of an hourglass shape. Late in pregnancy, she likely has the opposite shape, with a protruding abdomen. The increase in WHR provides a clear signal that she is not reproductively available, and thus most men should find her less attractive. The same trend is seen in women who have several children, resulting in a more tubular body shape; women who are older, resulting in a thickening midsection with the onset of menopause; or individuals who have high parasite loads, creating a protruding midsection. Each of these situations causes an increase in WHR, effectively communicating a decrease in reproductive potential. Men may consciously read these signals as merely a decrease in levels of attractiveness, but the signals are actually communicating information about underlying fertility and probability of reproductive success.

There has also been research examining the impact of variations of WHR on the ratings of male attractiveness. For males, a WHR of 0.9, which would create a tubular torso to hip appearance, is rated as the most attractive as well as the most common for Western males. This ratio would create the appearance of a solid midsection where the waist blends with the hips. However, variations around this ideal do not have the same impacts on the ratings of male attractiveness as similar variations for women. In fact, barring the very extremes, measurements of WHR seem to have very little overall impact on males' attractiveness. Similarly, WHR in males does not have the same predictive value for overall health or reproductive success as it does for women, which likely explains why it has not emerged as an important signal for attractiveness. Upon further examination, however, body proportions did emerge that were predictive of male attractiveness. However, rather than the WHR having an influence, it was shoulder-to-hip ratio (SHR) that matters.

Shoulder-to-Hip Ratio

SHR compares the circumference of the shoulder/chest to the circumference of the hips/waist. In males, SHR is more highly correlated with the ratings of attractiveness than WHR. Due to hormonal differences during puberty, adult males and females differ morphologically in the torso area. Broad, masculine shoulders are the results of male hormones while wide hips result from female hormone exposure. Thus, just as estrogen influences the development of hips in women, testosterone levels in puberty shape torso development in men.

In psychological research, men with a higher SHR (broad shoulders and narrow hips) were rated as more attractive by both men and women raters and provoked more feelings of jealousy by other males. These males were also more likely to be rated as physically and socially dominant. Further research examining men with a high SHR compared to men with low SHR found statistical differences in both behavior and physiology. Men with high SHR were more likely to have had sex earlier in their lives, were more likely to have had more sexual partners overall, had more testosterone in their systems, and were more likely to have cheated on their partners. Furthermore, men with high SHR tend to be biologically healthier. They are more likely to have healthier cardiovascular functioning and stronger immune systems as compared to men with lower SHR. This, again, supports the basic proposition that individuals use those traits that indicate increased levels of health to assess levels of attractiveness in others. Thus, the hormone levels that shape the body during development continue to have an impact on behavior later in life. When looking at women's SHR and behavior, none of these correlations was found.

Genetic influences have a definite effect on the development of SHR, which can be seen in the morphological differences that emerge during puberty. Additionally, environmental experiences can further impact development. Upper body workouts, diets, and clothing styles can accentuate SHR for males who are looking to improve their own perceived levels of attractiveness to others. Also, similar to women, the shape of the body again emerges as a more important indicator of attractiveness than overall BMI.

Leg-to-Trunk Ratio

Leg-to-trunk ratio (LTR) compares the length of the leg to the length of the torso. In the United States, women with long, toned legs are usually rated as more attractive. This may be one reason for the fashion trend of high heels. High heels tend to accentuate leg length, increase muscle tone, and enhance the buttock.

In 2011, Piotr Sorokowski and 38 colleagues from around the world examined LTR to see what actual impact it has on the level of attractiveness. They assessed over 3,000 participants from 27 nations and had them rate the attractiveness of male and female silhouettes that varied in leg length. Similar to the impact of averageness in facial attractiveness, they found that average or slightly longer than average legs were rated to be the most attractive. Although there is a stereotype of long-legged women being more attractive in Western cultures, for both males and females in this study, LTR closer to average was rated as the most attractive. These researchers found that excessively long or excessively short legs were perceived as less attractive than legs of average length across all cultures. The general pattern of preferences was consistent across cultures, but slight differences in the peak ideal did emerge. European, Canadian, and African participants rated ideal leg length as slightly longer than did Latin Americans. This could be a by-product of cultural influences, or typical norms for different ethnic groups.

In a follow-up study, Piotr Sorokowski and Agnieszka Sorokowska from the University of Wroclaw and Mara Mberira from the University of Namibia extended the data to include the Himba of northern Namibia, a semi-nomadic ethnic group. Similar to previous findings, extremely short or long leg length was rated as less attractive. However, a difference emerged in that the Himba preferred a relatively lower LTR for women (shorter than the average to above average ideal leg length rated as more attractive around the globe) and a relatively longer leg length for men.

Finally, Dr. Marco Bertamini, Dr. Christopher Byrne, and Dr. Kate M. Bennett from the University of Liverpool examined ideal LTR. They manipulated male and female images to range from relatively short-legged individuals to long-legged individuals, all within the possible norm. In line with the bulk of other research, they found that female bodies were rated as more attractive when they had average or slightly proportionally longer legs than the average woman.

Height

Average height varies from culture to culture and across generations. Over the past 100 years, research has demonstrated increases in average height around the world. Increases in average height stem from better prenatal care, better nutrition, and better health care for children. Increased height tends to be correlated with increased health and increased social benefits, at least to a point. In fact, we have a cultural adage that describes the most attractive men as tall, dark, and handsome. Psychological research examining female preferences of male height put this adage to the test. As predicted, research shows

---

Elongated Necks in Thailand

The process of creating elongated necks distinguishes the women of the Kayan tribe in Northern Thailand. The women in the Kayan tribe typically start the elongation process in childhood around age five using heavy neck rings. Young girls wear a few rings around their necks and then four or five more are added annually as they grow. The coils work by depressing the chest and shoulders of the girls and women. Thus, this creates the illusion of a head hovering over the rings without much actual stretching of the neck itself. The coils do not significantly lengthen the neck as commonly thought but simply compress the collarbone and upper ribs, which gives the appearance of a graceful giraffe neck. This process has not been found to be medically dangerous and removal of the rings will not cause the neck to flop or break as commonly believed. In fact, modern women of the Kayan tribe remove them regularly for cleaning and some choose not to sleep in the rings.

There are only about 500 members of the Kayan tribe remaining and they live in guarded villages on the northern border of Thailand. Anthropologists believe that the custom of women wearing rings to create the appearance of a long neck was meant to decrease their level of attractiveness in the eyes of other small-scale warring groups and also to mark the women as part of the Kayan tribe. Although this may have been the original purpose for the rings, the tradition has since had a drastically different effect, and the long graceful necks created by the process are now thought to be a sign of great beauty. There have also been claims that the rings help protect against and ward off tiger attacks. Tigers tend to attack victims at the neck and the rings symbolize protection of this vulnerable area. Groups with protected necks were thought to be less susceptible to drawing the eye of predators and thus the whole group was thought to be protected.

In the early 1990s, Kayan refugee camps were set up along the Thai border with long neck sections. This area has since become an independent tourist site. Word has spread about the beautiful long-necked women, and these women are able to make an income by posing with tourists to the region. Through such measures, the tribe relies on tourist revenue to support its survival.

Body Shape and Proportions 213

A young woman of the Kayan tribe in Thailand feeds her infant. Women of this tribe wear brass neck rings from childhood to create the appearance of an elongated, graceful neck. (Eldad Yitzhak/Shutterstock)

that taller men are considered to be more attractive by both males and females. Men who are above average in height are considered to be more attractive by females. Taller men are viewed as more mature, more physically dominant, and more powerful, and are rated as having increased abilities to care for offspring.

Biological evidence further supports the superiority of tall men. Research shows that increased height is correlated with better overall physical health and healthier genes. It also reveals that these taller individuals are likely a by-product of more stable rearing environments and have increased abilities to protect others. Increased height also tends to be correlated with increased strength and maturity.

Beyond physical benefits, research also shows that there is a positive correlation between male height and their reproductive success and that women rate them as more attractive. Conversely, women rate short men as relatively undesirable. When examining personal ads of actual males, taller men (over 6 feet) received more responses and were considered by women to be stronger and more athletic. Women even tend to give consideration to height when selecting sperm donors.

For males, the benefits of being taller even go beyond finding potential partners. Taller men actually tend to be more successful in everyday life. Taller men tend to have increased levels of confidence, make friends more easily, are promoted more quickly, are awarded more opportunities on the job, and thus have more experiences in their personal and professional lives. They tend to be preferred for jobs, are perceived as more effective, and are more likely to have more children than men who are shorter than average. Leaders of companies and individuals with higher socioeconomic status tend to be taller than average. Even in presidential elections, taller men have the advantage and are significantly more likely to be elected than shorter than average candidates. Taller men are perceived as stronger and more qualified, regardless of their actual experience and knowledge.

Research by Dr. Stulp and colleagues examined the life history of males and found that height is directly correlated with the number of surviving children. Interestingly, however, excessively tall individuals were actually at a disadvantage. Men who were taller but closer to average height were more likely to have the highest reproductive success. Men of average or just above average height were more likely to be married younger, and thus had more reproductive years to reproduce. Shorter and extremely tall men had decreased dating opportunities and tended to marry later. However, on online dating sites, women still show a preference for taller men. The same preference holds true in lab studies and questionnaire studies. Taller men are rated as more attractive during speed dating events, tend to date more attractive partners, and are more likely to eventually get married. Although average men are likely to have more children, taller men are more likely to be more dominant, have more leadership positions, have higher salaries and overall incomes, and are less likely to feel jealous of other men.

Height in women does not hold the same distinction as it does for men. Research does show a modest advantage for taller women. Specifically, taller women tend to have more children overall, and thus have higher reproductive success, and they also tend to have more economic success. Taller women tend to most frequently partner with men who are taller than themselves, so their partners tend to be taller men who likely experience all of the benefits that come with such distinction.

BREASTS AND BUTTOCKS

Although variations in shoulder-to-hip ratio do not have a significant impact on female attractiveness, breast size and shape do. Many men have a preoccupation with the female breast, so research focuses on what information is being communicated by the breasts. Basically, breasts provide indication of age, number of children, and pregnancy status. Young breasts tend to be rated as the most attractive. Young breasts are characterized by being firm and perky, and having small nipples. Men tend to rate larger breasts as more attractive, at least to a point, but symmetry matters. Large symmetrical breasts get the highest ratings but smaller breasts are likely to be more symmetrical, so they frequently are rated more highly than large, less symmetrical breasts.

While waist-to-hip ratio (WHR) emerges as a more reliable indicator of attractiveness, breast size has been found to play a role. Cross-culturally, men rate medium to large breasts as more attractive than smaller breasts. Furthermore, shape, symmetry, and areola color and size all influence attractiveness. Since clothing masks most of these characteristics, breast size is typically found to be the most important in male ratings. Breasts essentially provide visual cues of female fertility and genetic makeup. They are a natural by-product of the impact of sex hormones on body development during puberty and thus the size and shape of breasts communicate genetic health and maturity.

Although a widespread preference for larger breasts emerged in research, there were some cultural differences. In some areas of the world, smaller breasts were rated as just as attractive as larger breasts. Agnieszka Zelazniewicz and Boguslaw Pawlowski from the University of Wroclaw in Poland were interested in distinguishing between cultural influences on breast preferences. They discovered a connection between the sociosexual orientation of the society and breast preferences. Sociosexual orientation essentially describes the mating strategy of an individual or a culture. Individuals with a restricted sociosexual orientation tend to engage in more long-term and monogamous relationships. This strategy tends to arise in cultures where there are not very many available females. In such a society, to ensure a partner, the male must bend to the will of the females and remain faithful. An unrestricted sociosexual culture, on the other hand, indicates a society that engages in short-term mating and higher levels of promiscuity.

Dr. Zelazniewicz asked Polish men to rate 15 photographs of breasts for level of attractiveness. The male participants also filled out a sociosexual orientation measure to identify their level of restrictiveness. Men who scored high on sociosexual orientation, who endorsed engaging in short-term, low commitment sexual relationships (unrestricted men), found the largest breasts to be significantly more attractive than the restricted men. Restricted males, who endorsed engaging in long-term relationships, were less likely to emphasize breast size and more likely to emphasize emotional commitment. However, both groups of males preferred the average to above average breast size to smaller or excessively large breasts.

In a similar study from the University College London in the United Kingdom, Adrian Furnham and Viren Swami had British undergraduates rate breast size for level of attractiveness. In their review of the literature, they reveal that women with smaller breasts tend to be rated as more competent, ambitious, intelligent, moral, and modest by both men and women. In their study, they had both male and female undergraduate students rate the attractiveness of nine nude female silhouettes. The silhouettes varied by breast and buttocks size. The participants were asked to rate each figure for level of physical attractiveness. They found that breast size had an independent and significant impact on the ratings of attractiveness, with the smaller breast sizes being rated as more attractive and the largest breast size as least attractive, at least for these profile silhouettes. Since most studies tend to find medium or larger breasts to be more attractive, it may be the case that the larger breasts used in this study were unrealistic or that the profile view of the breasts made them appear excessively large. It could also be the case that a preference for larger breast size is not a universal preference and varies by culture or by individual. Buttock size did not have an impact on attractiveness ratings in this study.

A more specific examination of the attractiveness of breasts took areolar pigmentation into account. Barnaby Dixson, from the University of New South Wales, and colleagues had undergraduate students rate the attractiveness of photographs of women's torsos that showed very small, small, medium, and large breasts with varying levels of areolar pigmentation. Medium and large breasts were rated to be the most sexually attractive, most mature, and most nurturing by both genders. Furthermore, darkened areolar pigmentation served to increase the attractiveness ratings for the large breasts only. Small and medium breasts were rated more highly on attractiveness and reproductive health when they had lighter areolar pigmentation.

Males rate women with larger breasts as more attractive, but researchers were interested in examining whether this increased rating translated to actual behavioral differences. In a series of real-world experiments looking at men's actual behavior when confronted with women of varying breast size, men showed behavior differences that were directly contingent on breast size. For example, researcher Nicolas Guéguen found that, in nightclubs, women with larger breasts (simulated or real) were approached more frequently and more quickly. He also found that male drivers were significantly more likely to offer a ride to a female hitchhiker who had larger breasts. Likely due to increased attention for larger breasts, breast augmentation was the most common surgical cosmetic procedure in 2016, and has increased almost 40 percent since 2000. Of the almost 300,000 breast augmentation procedures over the last year, 100 percent were for women.

The beauty of the buttocks reveals them to be a similarly sexually dimorphic trait. The attractiveness of the female buttock ties in with WHR. Curvy hips and accentuated buttocks accentuate the hourglass shape for the female form. Thus, firm, rounded buttocks tend to add to a woman's level of attractiveness. Butt lifts and butt implants to create this rounded toned shape are up 200 percent since 2000 and are predominantly sought by women. For men, the ideal buttock depends on the width of his shoulders. Wide shoulders and narrow hips/buttocks tend to be rated as the most attractive for a male. Toned, strong, moderately sized buttocks tend to be rated as the most attractive for a man, particularly if he is sporting wide, muscular shoulders.

MUSCULARITY

Level of muscularity reveals health, fitness, and strength and thus likely has an impact on attractiveness. Defined muscles and toned bodies indicate youth, health, and wellness for both genders. However, since a man's ability to protect offspring and intimidate enemies is particularly attractive to women, researchers predict that muscularity is particularly important for males.

Interestingly, when asked about the importance of a man having large muscles, men rate the importance of having muscles more highly than do women. Men are also more critical of their own muscle mass than are women. Over the past few decades, men have become increasingly dissatisfied with their bodies, likely coinciding with the increased representation of extreme muscle mass in the media. Body builders, professional athletes, and male models have become progressively more muscular and more in the public eye over the last few generations. Even our children's toys now exhibit male figures with muscle mass to rival or surpass the top body builders in the world.

Due to this extreme ideal, men are being subjected to the same pressures regarding body image that have plagued women over the last century. Current research shows that men who lack muscle mass believe that they are not desirable to women. Shaun Filiault from the University of South Australia examined men's perceptions of their own bodies and the influence on their sex lives. Dr. Filiault found that, throughout his sample, men were frequently dissatisfied with their own levels of muscle and fat. Lack of muscle tended to impair self-esteem, increase risk of depression, and create a preoccupation with one's appearance. Finally, lack of muscle had an impact on a man's sense of sexual ability and masculinity. Increased levels of fat, interestingly, did not seem to impact a man's sense of masculinity. Increased levels of fat did have a negative effect on self-esteem and peers bullied overweight males more.

So, regardless of actual female preferences, muscularity seems to be important for a man's psychological health and self-esteem. Muscle mass for men in a body-conscious society can positively or negatively impact body image, which can have compounding effects in other areas of their lives.

INGUINAL CREASE

The inguinal crease is the v-shaped indentation between the front of the hip and the lower abdominals. In an online survey, the inguinal crease emerged as among the top five features that women rate as the most physically attractive on a man. This crease is created from a ligament that spans from the hipbone to the pelvic bone. Although everyone has this ligament, it is not externally visible unless the individual has very low body fat and well-defined abdominal and core muscles. The combination of less than 5 percent body fat and well-developed surrounding muscles is what gives it definition. Since low abdominal fat is required to expose the inguinal crease, individuals with a visible line likely have a narrow waist and hips, increasing their shoulder-to-waist ratio. This may be one reason that the inguinal crease increases the attractiveness of the male form more than it does the female form.

For those seeking to enhance their inguinal crease, men's health magazines encourage diet and exercise. Abdominal exercises that target core muscles, such as the transversus abdominis, are needed to enhance the crease. This muscle lies along the side of the torso and is a deep core muscle that stabilizes the torso. Along its lower edge, it attaches to the inguinal ligament; so enhancing the size of this muscle helps define the crevice along its lower edge. Enhancing the size of this muscle while reducing excess abdominal fat culminates in a more defined and sexy crease. The inguinal crease also tends to trap body scent and, if up close and personal, can increase or decrease attraction based on pheromone attraction.

MALE GENITALIA

Situated below the inguinal crease are the male genitalia. Similar to female breast size and shape, male genitalia come in various shapes and sizes.

Given the natural variation, many men want to know, does size matter? Fortunately for questioning men, there is no empirical evidence that women prefer a larger penis. In Cameroon, Dr. Barnaby Dixson asked women to rate photographs of male figures with a nonerect penis. The smallest (reduced to 78 percent of the normal length) and largest (43 percent longer than average) were rated as less attractive than the three more average examples. Based on this study, having an average penis is likely most beneficial for attractiveness, but an abnormally large penis does not confer an advantage.

Most other research finds that women vary in their personal preferences and an overarching preference for a larger penis has not been empirical validated. Although some women may endorse a preference for a larger-than-average penis, no consistent preference has been found that spans across all women. Thus, ideal penis size seems to be a matter of personal preference. These findings may be reassuring to the American male who is concerned about his own anatomy and whose sexual confidence revolves around penis size.

Currently, as a result of the insecurities of men in the United States about penis size, interventions and contraptions are marketed that claim to enlarge one's penis. These include things such as pills, creams, pumps, and surgical interventions. Unfortunately, many of these methods are ineffective or cause more harm than good. For example, in an attempt to help insecure men enlarge their penis, liquid injectable silicone has been used for penile augmentation. Unfortunately, as reviewed by Dr. Jonathan Silberstein and colleagues from the University of California and Alvarado Hospital, this procedure frequently has poor long-term outcomes, leading to the need for surgery to correct complications. Complications can include pain, swelling, abscesses, allergic reactions, deformity of the penis, scarring, ulcers, and sexual dysfunction. Although thousands of men have undergone injection, many of these have led to the need for surgical removal of the silicone.

FEET

Foot size is another trait that may impact the level of attractiveness for men and women. Since foot size is a sexually dimorphic trait, small feet in women and large feet in men may enhance the distinction between the genders, leading to increases in attractiveness ratings.

A study by Daniel Fessler from the University of California and colleagues from eight other schools examined the impact foot size had on the ratings of attractiveness. They used line drawings of males and females that varied only by foot size and asked participants to rate which one was the most sexually attractive or beautiful/handsome and the one that was the least sexually attractive or beautiful/handsome. They found that in nine different cultures,

Foot Binding in China

Since foot size is a sexually dimorphic trait, smaller feet in women enhance femininity and can contribute to their overall attractiveness. An extreme example of the quest for small feet can be found in the now extinct Chinese tradition of foot binding. For hundreds of years, foot binding was practiced throughout China. In childhood, before a girl's foot grew to full size, the four smaller toes on each foot were broken and tightly bound against the bottom of the foot, creating a triangular shape. Tight wrappings forced the foot to be bent double and prevented the foot from growing to a normal size.

The resulting small foot was heralded as an ideal of feminine beauty and became a status symbol among the elite. A 3-inch foot was the most ideal and was referred to as the golden lotus. Four-inch feet were acceptable and called silver lotuses, but anything bigger were referred to as an iron lotus and were deemed unattractive. Women with the

The bound feet of an elderly Chinese woman from Yunnan Province in China. Tight wrappings in childhood stunted foot growth and impeded mobility, but the tiny feet were heralded as extremely beautiful and feminine. (Torsten Stahlberg/iStockphoto.com)

smallest feet were rated as the most attractive and were most likely to secure a high bride price and a good marriage.

Although the tradition of foot binding began among the wealthy, it became widespread even across poorer populations. Anthropologists note that the process increased the women's ratings of beauty and desirability, but it also had the effect of encouraging girls to sit still to be more productive in textile production. Young Chinese women were expected to spin, sew, weave, and stitch, and their feet tended to be bound at the same age when girls were ready to start learning to do such work. As the Chinese culture has changed and young girls do not as commonly produce textiles, the practice of foot binding has decreased as well.

Although foot binding may seem bizarre to European and American women, our history holds a somewhat comparable practice of corseting during the Victorian era. The binding corsets created the ideal curvaceous figure but tended to change the shape of the ribs and shifted and impeded growth of internal organs. Corsets left women short of breath and unable to stand for long periods of time without support. Furthermore, use of high-heel shoes is a fashion choice that lightly mirrors the food binding in China in that this fashion choice minimizes the appearance of foot size for women. These shoes also damage women's arches, and cause foot and back problems with long-term use.

small foot size was generally rated as more attractive for females and average foot size was preferred for males. The smallest footed female figure was rated the most attractive and the two largest footed female figures were rated the least attractive. For men, the average foot size was rated as most attractive and the largest and smallest male-footed figures were rated as the least attractive. Small feet may indicate youthfulness (children have very small feet) and enhanced femininity, explaining the preference for smaller feet for women across these cultures. For males, the proportionate feet were rated as most attractive, aligning with the idea that averageness is beautiful.

HORMONAL INFLUENCES ON ATTRACTION

The Allure of Ovulation

As discussed throughout this text, women have been found to exhibit different preferences throughout different phases of ovulation. During their most fertile phase (around two weeks after the last period), women rate more masculine faces and more masculine bodies to be the most attractive. For example, in a study with Caucasian women, when participants rated photographs for level of attractiveness, their preferences varied predictably over their cycles. When they were the most fertile, they preferred darker Caucasian male faces. They also preferred more masculine, prominent chins and thicker eyebrows than when in a nonfertile phase of the cycle. All of these traits emphasize the sexually dimorphic features between the sexes and may indicate a strong potential mate.

The effects of ovulation do not only affect female preferences. Males, similarly, are affected by a female's level of fertility and rate females as most attractive when the women are in their most fertile phase. These females are most likely showing the most interest and sending covert signals of attraction. For example, women who are ovulating produce pheromones called copulins. When men smell these scents, their testosterone level goes up, causing the men to produce body odors that repel women who are not ovulating. Thus, there is an entire game of tug of war at the molecular level of which most are not even consciously aware.

Although most research on attraction focuses on physical and overt characteristics (such as wealth and status), there are covert signals that also come into play when seeking, developing, and maintaining relationships. Romantic desires, for example, can be influenced by subtle scents, particularly those related to levels of fertility. These scents play a particularly important role for men because, while men's level of fertility remains relatively stable over time, women's levels change drastically over the course of a month. It would make sense, therefore, that men should be most attractive to a woman at the point of highest fertility, or right before or during ovulation. Because women do not display overt signals of ovulation (unlike the bright coloration of the female baboon posterior when fertile), research has turned to the study of covert signals. Findings of such signals include slight changes in body odors, shifts in vocal pitch, and changes in skin tone during ovulation. When asked to rate women over the course of their cycle, the highest ratings of attractiveness were awarded when women were in their most fertile phase. Furthermore, these cues of heightened fertility also influenced the testosterone levels of the men, making them more motivated and interested in attraction.

Saul L. Miller and John K. Maner from Florida State University examined the effects that the menstrual cycle has on male participants' ratings of attractiveness for a female confederate. As expected, the single male participants rated the female as more attractive during the most fertile part of her cycle. They also rated her as more intelligent during peak fertility. A similar change in ratings did not occur when female participants were asked to rate the female confederate's level of attractiveness. Interestingly, men in committed relationships actually rated the woman as less attractive during her most fertile phase. Thus female fertility cues may evoke differential responses depending on the audience. Single males show increases in their level of attraction, while committed males suffer a decrease in attraction. This may be one evolved mechanism to help maintain long-term relationships even in a sea of possible alternatives. Research examining the males' behavior toward their own long-term partners during times of increased fertility would likely reveal increased levels of attraction and increased mate-guarding against potential threats, further solidifying the commitment.

Men find female scent near ovulation as more attractive than when she is in less fertile phases of her reproductive cycle. This phenomenon has also been observed in hamsters, dogs, cows, sheep, and other primates, such as rhesus macaques. Those females who are naturally producing or have been rubbed with estrous scents attract male sexual advances significantly more than those who are not in the fertile phase. Dr. Kelly A. Gildersleeve and colleagues from the University of California and California State University had women collect samples of their scent twice during their cycle: once in the most fertile phase just before ovulation and once during a nonfertile phase. As predicted, the scent samples taken near or during ovulation were rated as significantly more attractive than the scent samples taken during the nonfertile phase. Additionally, the closer the woman was to ovulation when she collected the fertile sample, the more attractive it was to male raters.

The fact that men find women more attractive during the most fertile days of women's cycle influences male behavior in relationships. Males tend to be more attentive and interested in their partners during the females' most fertile phase. They spend more time with their partner, engage in more mate-guarding, and display more jealousy. Exotic dancers not using hormonal contraception earn considerably more tips near ovulation. Men who meet a woman during her more fertile phase tend to like her more and engage in riskier behaviors than men who interact with the same women on a low-fertile day.

When examining these interactions, it is difficult to establish whether the female acts differently when most fertile, which then influences the male behavior, or if the male behavior is entirely due to hormone exposure. Given the breadth of study, there is likely a combination of effects happening. Based on Dr. Gildersleeve's scent study, men could perceive fertility just based on scent, without even interacting with the woman. In the case of the exotic dancers and relationships, the woman's behavior may compound the effects. For example, during high fertility, women may be more interested in sex, may show more approach behaviors, may show more skin, or may make more eye contact.

Dr. Martie G. Haselton and colleagues from the Center for Behavior Evolution and Culture at the University of California and from University of Wisconsin examined changes in women's grooming habits and clothing choices throughout their ovulatory cycle. When selecting from photographs taken at different points in the cycle, men were more likely to choose the photos taken during the fertile window as most attractive. These photographs showed evidence of more grooming and ornamentation and of measures taken by the women to look more attractive.

Facial photographs during the fertile phase tend to be rated as more attractive, scents during the fertile phase are rated as more attractive, and now grooming habits have been demonstrated to shift as well. Even women's likelihood of moving around and volunteering for social activities increases during ovulation. Furthermore, women's reported interest in sex peaks during the most fertile phase and they report more interest in flirting and social exploration. During this phase, men tend to be more possessive, jealous, and attentive to their partners. In Dr. Haselton's study, the photographs of the women during their fertile phase showed evidence of more ornamentation of dress. The women wore more fashionable or nicer clothes, wore tanks with lace trim, wore skirts rather than pants, added decorative scarves, and showed more skin overall.

Men rate their partners as more attractive during the fertile phase of the cycle than during the luteal phase or during hormonal contraceptive use. Men also rated themselves as more attractive when their partner was ovulating and free from hormonal contraceptives. This change in males' self-perceptions may be due to finding their partner more attractive or due to their partner's increased interest in sexual activity. It is likely a combination of scent and behavior that accounts for changes in interest and attraction throughout the reproductive cycle. Men may find women more attractive and women may present themselves more attractively.

Birth control that alters hormones can influence attractiveness ratings and skew perceptions. Not only do the hormonal influences alter the preferences of the women taking them but they may alter the perceptions of the surrounding males as well. Women taking hormone-mediated birth control tend to be viewed as less attractive to males. Women not on hormonal birth control are most attractive to males during their more fertile phase. On hormonal birth control, women do not have a most fertile phase because the hormones trick the woman's body into thinking it is already pregnant.

The Impact of Female Hormones on Male Attractiveness

Victor S. Johnston and colleagues from New Mexico State University, the University of New Mexico, and Karl Grammer from Ludwig-Boltzmann Institute for Urban Ethology in Vienna examined the impact of hormones on male attractiveness. Surprisingly, however, male hormones were not the trait under examination. Instead, fluctuations in female hormones throughout the month play a role in male attractiveness. As discussed in the last section, female hormone levels impact male ratings of female attractiveness but Johnston and colleagues demonstrate that fluctuations in female hormones similarly affect female ratings of male attractiveness.

Early research studies on the ideal male face tended to contradict each other. Some report that women prefer more masculine faces and others conclude that more feminized features are preferred. In Johnston's study, he examined whether the menstrual phase could help explain the difference in the findings. Female participants rated male faces at two different points in their menstrual cycles. Using male and female facial images, participants were asked to select faces that fit a variety of descriptors. Among these 15 descriptions were a masculine-looking male, a feminine-looking female, an intelligent-looking male, an intelligent-looking female, a good father, and a good mother. For each face selected, participants rated the face on 20 different attributes such as physically attractive, sexually exciting, intelligent, trustworthy, good parent for a child, threatening, manipulative, and sensitive. Two weeks later, women returned and completed the same task in order to see changes over the phases of their cycle. Upon analysis, 29 of the female participants were in the limited window of peak fertility when rating the male faces. These women were more likely to select a more masculine face as most attractive than when they were not most fertile. Furthermore, out of all of the faces and descriptors they rated, only the attractive male face preference changed between the two sessions. Women during a more fertile phase report preference for a more masculine partner. Although preferences for all women shifted in line with their phase, females who were more feminine had a greater change in preference than more androgynous women.

The more masculine faces, which tended to be rated as more attractive, had more overt markers of testosterone exposure. These include features such as a strong jaw, brow ridges, and pronounced cheekbones. The extent of masculinization of the face positively impacted the ratings of attractiveness, at least to a point. Hypermasculinity was associated with lower scores. So, there seems to be a balanced level of testosterone that maximizes attractiveness and higher levels tip the scales to dominant, unfriendly, threatening, and manipulative ratings. The optimum balance, however, may be a moving target, depending on the woman's phase of fertility. Attraction to testosterone is mediated by the presence of estrogen.

Major Histocompatibility Complex and the Scent of Attraction

The major histocompatibility complex (MHC) is a set of genes that the body's immune system uses to discriminate pathogens. In lab animals, such as mice, MHC can be detected through scent and either attracts or repels mating behavior. This ability to unconsciously detect MHC through scent has also been demonstrated in birds, fish, some reptiles, and humans. Individuals tend to prefer others who have genes that are varied from their own. In general, when examined, married couples tend to have more dissimilar genes than would be expected by chance. This likely is a natural defense against incest since genetically related family members have very similar genes, making them less sexually attractive. Furthermore, too little overlap in MHC is similarly unattractive. There seems to be an ideal balance of enough dissimilarity to guarantee genetic variability in offspring but enough similarity to ensure common experiences and goals.

Dr. Christine E. Garver-Apgar and colleagues from the Department of Psychology and the Department of Biology at the University of New Mexico analyzed romantically involved couples to examine the MHC pairings. They predicted that couples who had MHC that were too similar would be less sexually responsive, that they would more likely have sex with other people, and that the women would be most attracted to other men during their most fertile phase. After analyzing 48 couples, they found that women did report less sexual attraction to their partner if their MHC genes were similar. Men in these situations also reported that their partner tended to be low in sexual responsiveness. While overall relationship satisfaction varied by couples, satisfaction with the sexual aspect of the relationship was significantly correlated with the level of shared MHC. Having higher levels of shared MHC was also directly correlated with having sex with other people while in the relationship. Women in these relationships were also most attracted to other men during the most fertile phase of their cycles. Although MHC sharing did not seem to alter men's interest and arousal, it did alter their ratings of how interested their partner was in the relationship. So, men were able to perceive the decrease in interest and arousal that women showed when they shared too many MHC markers. Women in relationships with too many shared MHC genes were more likely to reject their partner's efforts to engage in sex, especially when the women were in their most fertile phase, and women experienced fewer orgasms with partners as the level of MHC shared increased.

Although MHC is not a conscious marker of attraction, it does seem to influence women's tendencies, behaviors, and interests throughout the fertility cycle. High similarity in MHC causes women to show less interest and less engagement in the sexual aspect of their relationships, particularly during their most fertile phase. At the same time, they are more likely to engage in sexual activities with another person during this phase, particularly one who is dissimilar on MHC markers. In other animals, this unconscious analysis seems to occur through scent and it may function similarly in

Hormonal Influences on Attraction 227

humans. The natural chemistry and physical arousal that occur when two people are physically close may be heightened if they have dissimilar genes and immune systems. The benefit of mating with someone with a dissimilar immune system is that offspring would have more genetic diversity and be more likely to survive in a changing environment.

In additional research, Randy Thornhill, Steven Gangestad, and colleagues from the University of New Mexico completed similar studies on the MHC. In their foundational study, researchers asked men and women to rate the attractiveness of the scent of T-shirts worn by the opposite sex for two nights. Men selected T-shirts worn by women with dissimilar MHC who were in the most fertile phases of the menstrual cycle as the most attractive to the nose. These women were also those who were rated as the most physically attractive by other viewers, showing that physical appearance and scent are correlated.

Interestingly, for women using hormonal contraceptives, ratings were not correlated with MHC dissimilarity or with facial attractiveness, though they did prefer men who demonstrated MHC heterozygosity, particularly when making ratings during their most fertile phase. Women using hormonal contraceptives frequently preferred males with similar MHC genes. Hormonal contraceptives trick the body into thinking that it is already pregnant. During times of pregnancy, it would make sense for a woman to seek family and protection. Thus, in those times, she should be drawn toward those who are genetically similar to her. When not pregnant (or not on hormonal contraceptives), it makes evolutionary sense for her to be attracted and drawn toward those who are dissimilar (and thus a better choice for a reproductive partner). A potential negative effect of hormonal contraception may be that women are attracted to males with high MHC similarity rather than those with diverse genes. Problems may emerge when the woman stops taking the hormone. Once she is no longer being influenced by the hormonal contraceptive that influences her preference for genetically similar individuals, she may no longer be attracted to, or may even be repulsed by, her partner. If, for instance, a couple stops using contraception in order to conceive a child, the woman may no longer being attracted to the man's scent or sexually attracted to her partner during her fertile phase.

On the contrary, women who are taking hormonal contraceptives during a relationship may be less attractive to their partners. In an empirical study, researchers demonstrated that exotic dancers who were taking hormonal contraceptives earned about 50 percent less in tips than dancers who were not taking a contraceptive. Taking a hormonal contraceptive decreased the sexual attraction and interest levels of the men.

Since the effects of MHC are at the unconscious level, they likely only have a subtle impact on behavior. When meeting someone new, MHC may spark interest, direct attention, or provide a sense of instant chemistry. Once a relationship is established, however, ending contraceptive use or beginning contraceptive use will not necessarily have a substantial effect. The relationship is hopefully well enough established and the individuals so committed that the slight difference in smell will not have a noticeable effect. Once two individuals are committed to one another, other factors such as intimacy, familiarity, and commitment may be valued more highly than scent.

Given the importance of body scent on attraction, perfumes, colognes, shower gels, body wash, soap, shampoos, and lotions may be interfering with our natural reproductive processes and choices. We mask our scents so intensely that it may be difficult to find that individual to whom we are naturally attracted. Couples who mask their scent may enter into relationships that are not reproductively cohesive. For example, couples with highly similar MHC are more likely to have difficulty conceiving a child, experience greater risk for spontaneous abortions, and have offspring who are underweight and premature. The woman's body seems to recognize the ineffectiveness of the lack of genetic variation, leading to natural termination. Thus, smell is important for the success of a relationship, for the individuals involved, and for their potential offspring.

If attempting to initiate a long-term relationship, it may be worthwhile to eliminate fragrances and hormonal contraceptive use. Finding a partner who is attracted to one's natural scent may increase the likelihood of success of the relationship and potential future offspring. In previous eras, free from soap and deodorants, individuals had an easier time smelling and assessing their choices. Today, we have to get up close and personal to even catch a whiff of our partner, and even then it may not be an accurate representation of their natural smell.

That being said, there is currently no evidence that perfumes laced with pheromones actually work to attract a mate. Less intense odors tend to be rated as more attractive, so science would recommend that one use caution when applying fragrances. Scent is a subtle cue and it can quickly become overpowering. Furthermore, one's natural scent may be the most alluring, so if one does use perfume or cologne, keeping it light enough to not cover natural scents may be the most beneficial. In a romantic match, individuals tend to like the way their partners smell naturally, even if the scents may be a turnoff to someone else.

However, body scent can also impact an individual's own sense of confidence. Research by Dr. Roberts and colleagues from Liverpool University demonstrated that increases in body odor decreased males' self-confidence and attractiveness ratings. Men were rated for level of attractiveness at the onset of the study. Then half of the men were placed in a placebo group and were instructed to use a spray on deodorant for two days that was actually inactive. The other half was instructed to use a normal spray-on deodorant.

After two days, each participant was instructed to create a video where they imagined they were introducing themselves to an attractive woman. When women rated the videos, those men who had been using the inactive deodorant were rated as less confident and less attractive than they had been rated at the beginning of the study. Thus the body odor that was unsuppressed by the fake deodorant actually impacted the way the men presented themselves to others. Although women could not smell them via the videotapes, the drop in confidence was evidenced in their behavior. Thus, the effective use of perfumes and colognes may be that they bolster confidence in addition to communicating an attractive smell.

In an additional study, women were asked to rate the attractiveness of photographs of men. While they were rating, the women were exposed to either pleasant or unpleasant odors. Women gave ratings of higher attractiveness to the photographs they were rating while they were being exposed to the pleasant smells and they gave the lowest ratings to the photos of individuals they saw while being smelling the most unpleasant odors.

Overall, smells can increase or decrease blood pressure, motivate people to protect others, spend money, or be irritable, or increase or decrease mood. If individuals are attracted to pheromones and body scent, it may be more beneficial to not cover up natural body scent when seeking a relationship. Finding an individual who is attracted to one's natural smell may result in happier, more sexually active, and more productive reproductive relationships.

Scent is connected to memory and emotion. Kissing may have developed from the act of sniffing one's partner. Being up close and personal with a partner gives the olfactory bulbs more information about their natural scent, even if they are attempting to mask it with perfumes or colognes. The kiss engages several senses and is a great way to measure the level of attraction. An individual gets a sense of personality, scent, and sensuality through a kiss. Over 60 percent of women report that they would end a relationship with another person if there is a bad first kiss. Frequency of kissing in a relationship is directly related to relationship satisfaction. A 10-second (or longer!) kiss every day is correlated with significantly less stress and rates of depression within a relationship.

The attractiveness of smells tends to vary from individual to individual. Smells that entice one person may repulse another and this extremity extends to body odor. Furthermore, body odor is more likely to repel a mate than to seal the deal. Psychological stress causes odor that differs from natural sweat. This means that women can detect a stressed man.

Body scent does not have to be conscious to be effective. Men and women may not realize it is the scent that is sparking their sexual desire. The steroids that may be emitted from a partner can stimulate the nerves in the nose and case an entire chemical reaction without conscious awareness. But, the absence of unpleasant body odor seems to be more influential than the presence of pleasant odors and we are more attracted to the odor (via worn T-shirts) of those with a dissimilar immune system to our own. Ultimately, women tend to find a male's scent even more influential than his physical features.

Hormones' Impact on Vocal Quality and Attractiveness

Voice quality is another sexually dimorphic trait that tends to vary predictably between the sexes. Females' voices tend to be higher pitched and males' voices tend to have a lower pitch. According to evolutionary psychologist David Buss, a man's lower voice tends to be particularly attractive to women. A deep voice is not only an indicator of sexual maturity but also correlated with a larger overall body size. A deep voice is indicative of healthy genes and of a more dominant, more powerful partner. Similar results were found when examining male preferences for female voice quality. In research studies, men rate women with higher-pitched voices as the most attractive. However, for both genders, pitch is most influential in initial attraction and less important once a relationship has been established.

During the most fertile phase of a woman's cycle (around two weeks after her last period), deeper, more masculine voices are the most attractive. Thus, in research, a deep voice is more correlated with being attracted to short-term partners and not rated to be as important when women are considering a long-term relationship (e.g., it is attractive but not necessary). Furthermore, during peak fertility, men rated women's voices as the most attractive. At this point, women's voices were found to be the most seductive and men reported that their skin would tingle when listening to the speech of a female confederate.

As with other physical features, hormone levels impact voice quality during development, so this overt feature provides insight into underlying physiological functioning. Since hormone exposure has widespread influences throughout the body, voice pitch should correlate with other features. Researcher Dr. Wade demonstrates that men with deep voices also tend to have broader shoulders and a higher shoulder-to-hip ratio, both features that are also attractive to women. Women with high-pitched voices tend to have lower waist-to-hip ratios, another attractive physical feature. For both genders, an attractive voice is correlated with the symmetry of the face and body. All of these features taken together provide strong evidence that the individual is healthy and stable, and will be successful reproductively. Because hormones influence voice pitch, it makes sense that there would be widespread indicators of this internal quality.

Estrogen raises the pitch of the voice. During ovulation, women's voices are higher pitched than during other times of the cycle. When asked to rate voice quality and attractiveness, men consistently chose higher pitches as most attractive, although unconsciously, they were choosing the voice that signaled greatest fertility.

PSYCHOLOGICAL TRAITS

All of the physical traits discussed throughout this section provide insight into what humans tend to find attractive in themselves and others. Strong, healthy, youthful bodies are clearly the most attractive when encountering a new potential partner. However, when rating people whom we know personally, judgments tend to become skewed. Once an individual is known and liked, it is more difficult to objectively rate the level of attractiveness of individual features. Personality characteristics, shared intimacy, intelligence levels, common history, and personal connections can impact ratings, even if one is specifically asking about physical attractiveness.

Personality, Intimacy, and Length of Relationships

Many traits beyond physical features can influence the ratings and perceptions of attractiveness. Having ideal physical traits but an annoying or aggressive personality can radically alter how one is perceived. Outgoing, interesting, engaging, and kind people tend to be perceived as more attractive than their actual physical features would predict. Extraverted, friendly individuals are rated as more attractive, more intelligent, more successful, and more capable even if they are physically only average.

Familiarity, mutual physical attraction, shared interests and experiences, and attractiveness increase intimacy, or a sense of closeness. Intimacy positively impacts relationship satisfaction. Perceiving someone as familiar aids in comfort and likability. Familiar people remind us of those we already spend time with, and we are more likely to engage in relationships with those who are familiar. Having similarities in attitude or interests makes the time spent together more enjoyable and more fulfilling. It also means there is less likelihood of conflict. Trust tends to emerge from intimacy as well, and we are more likely to confide in, seek support, and engage in relationships with those whom we trust. Once intimacy has developed, ratings of physical attractiveness become less objective and individuals tend to appear more and more attractive even if their physical traits remain unchanged.

Having a long-term relationship with someone increases familiarity, shared experiences, and intimacy. The development of these traits can increase the ratings of attractiveness that extend beyond mere physical traits. Judging a book by its cover may only suffice early in meeting someone new. Personality, shared experiences, and intimate connections prove to be more informative about overall attractiveness. Ultimately, subjective attractiveness becomes more influential than objective attractiveness once a relationship has been established.

Conclusion

Research exploring the science behind human attractiveness has revealed that there are individual and cumulative contributions to overall attractiveness. Although culture and media can help shape decisions, unconscious processing largely drives judgments of attractiveness, and many choices are made with respect to overall health and reproductive potential. Attractiveness is revealed in the quality, color, and placement of hair; the quality, color, and texture of skin; and the shape, placement, and structure of facial features. Traits that exhibit genetic health and developmental stability lead to higher ratings of attractiveness, and a more attractive physical appearance reveals gender-specific hormone exposure during development. Body shape and specific body features reveal health and reproductive potential as well as personal grooming habits. The attractiveness of scent as well as other hormonal influences helps ensure successful reproduction and long-term interaction. Finally, psychological effects contribute to relationship satisfaction and continued interest and commitment in partners and offspring. The biological, psychological, and social effects on the assessment of attractiveness make its study both an objective and subjective affair.

Glossary

Affect: An emotion or desire. For example, increases in positive affect would include greater feelings of happiness or enjoyment while negative affect includes such emotions as sadness, anger, or disappointment.

Altruism: Acting to benefit others rather than to benefit oneself.

Androgynous: A description of an individual who is neither extremely feminine nor masculine but expresses both masculine and feminine traits.

Anorexia: A psychological disorder characterized by disordered eating and distorted body image. Individuals with anorexia tend to restrict food intake and may purge, exercise excessively, or use laxatives to maintain an extremely low body weight. Individuals may feel fat or fear becoming fat even though they are underweight.

Anthropology: The study of human societies in the past and present. Anthropology considers social, biological, and cultural influences on one's development and behavior.

Attractiveness: The extent to which one's physical or behavioral characteristics are pleasing and appealing to others.

Averageness: In attractiveness research, averageness is the result of combining multiple facial images to create a standard image, devoid of the extremes of any one face. Once these images have been averaged, the resulting facial image tends to be rated as more attractive than the contributing individual faces, likely because imperfections and asymmetries are eliminated.

Beauty standards: Culturally constructed ideals of which traits are the most attractive.

Big Five personality traits: A model of personality that describes personality along five independent dimensions: openness to new experiences, conscientiousness, extraversion, agreeableness, and neuroticism.

Blepharoplasty: A surgical procedure used to reshape the eyes.

Body consciousness: Awareness of one's own physical appearance and excessive concern about how one may be perceived by others.

Body Image: How one thinks and feels about one's own body.

Body mass index (BMI): The ratio of one's weight and height (weight/height). This calculation is used to determine whether an individual is underweight, average, overweight, or obese.

Body modification: Intentionally and permanently changing one's physical appearance. Common types include piercings, tattoos, and plastic surgery.

Body symmetry: For human attractiveness, body symmetry refers to the degree to which the right side of the body matches the left side of the body. Higher symmetry is correlated with higher ratings of attractiveness.

Bulimia: An eating disorder characterized by binge eating followed by purging, through either vomiting or use of laxatives.

Cardiovascular health: Refers to the health of one's heart and blood vessels.

Cognitive script: A mental representation of how an experience should occur based on previous experience.

Complexion: The color, texture, and evenness of the skin, particularly for the face.

Confederate: An individual who is part of the research team who plays a role in the study to elicit a response from the participants. Participants in the study think the confederate is another participant in the study.

Conscientiousness: The Big Five personality dimension that indicates an individual's level of orderliness, dedication, and attention to detail.

Correlation: The extent of the connection between two variables.

Cosmetic surgery: A type of plastic surgery that aims to improve the attractiveness of features of the face or body.

Cosmetics: Products, such as lotions, creams, or makeup, used to alter or enhance the appearance of the face or body.

Depression: A mood disorder characterized by prolonged periods of sadness and loss of interest in activities.

Double eyelid: An eyelid that contains a crease in the flap of skin over the eye.

Elongated necks: A characteristic of the Kayan people who use neck coils to depress the collarbone and upper ribs to create the appearance of a longer neck.

Glossary

Estrogen: The female sex hormone that causes feminization of female primary and secondary sex characteristics.

External locus of control: The degree to which individuals believe that external forces have control over the outcome of a situation. Those with an external locus of control do not believe they can affect the outcome of a situation because the control is external to themselves and thus they are less likely to try to make changes.

Extraversion: The Big Five personality dimension that indicates an individual's level of energy and interest in engaging with others. Extraverts tend to be outgoing, be sociable, and become more energetic when engaging with others.

Evolution: The process of passing on adaptive genetic traits through generations via the process of natural selection.

Evolutionary perspective: The perspective that human behaviors and mental processes are a by-product of natural selection. Those traits that promote survival and successful reproduction have been passed on through generations, shaping the genetic makeup of the species.

Evolutionary psychology: The examination of psychological traits with respect to our evolutionary history. This is a theoretical approach that explores the function of modern behaviors and mental processes.

Facial symmetry: The degree to which the right side of the face matches the left side of the face. More symmetrical faces tend to be rated as more attractive.

Feminine: Possessing the traits or features that are typically associated with females.

Fertility: The ability to reproduce.

Foot binding: The process of inhibiting growth of the foot through tight bindings and breaking of the bones. This process was historically used throughout China and was correlated with enhanced ratings of female attractiveness.

Fundamental attribution bias: Our tendency to explain others' behaviors as a by-product of their disposition or personality, without considering the external circumstances that may be influencing them.

Genotype: The genes that a specific individual possesses in his or her DNA.

Golden ratio: The specific ratio of length to width that tends to be rated as the most pleasing to the human eye. Also known as the divine proportion, this ratio naturally occurs in phenomenon such as flower petals, shells, hurricanes, faces, fingers, body proportions, and DNA molecules, and is mimicked in the most famous works of architecture and art.

Halo effect: When an individual possesses one positive trait, such as a high level of attractiveness, he or she is perceived to have other positive traits as well.

Heterosexual: When an individual is sexually attracted to individuals of the opposite sex.

Homosexual: When an individual is sexually attracted to individuals of the same sex.

Homozygous recessive trait: Traits that are only expressed if the genotype contains only alleles of the recessive form. For example, two alleles for blue eyes are typically needed for an individual to have blue eyes.

Immune system: The body system that protects the body from pathogens or other foreign substances.

Infidelity: Being unfaithful to a partner.

Internal locus of control: The degree to which individuals believe they have control over the outcome of a situation. Those with an internal locus of control believe they have the power to change the situation, achieve the outcome they want, and solve problems, and thus, they are more likely to confront problems or try to improve relationships.

Intimacy: A sense of physical, psychological, spiritual, or emotional closeness with another person.

Iris: The colored part of the eye.

Laser hair removal: A medical technique used to damage hair follicles to remove unwanted body hair and prevent its future growth.

Leg-to-trunk ratio: The comparison of the length of the leg to the length of the torso. Average to slightly longer than average legs tend to be rated as more attractive.

Limbal ring: The darker ring around the iris that is more prominent in children and young adults.

Long-term mating strategy: The intention to commit to another person and continue the relationship into the future. For example, marriage is a long-term mating strategy that signals an intention to stay together for the long-term.

Lumbersexual: A man who attempts to look more masculine through emulating the characteristics of a lumberjack, such as growing a beard or wearing rugged clothing.

Major histocompatibility complex (MHC): An aspect of the immune system that allows for identification of foreign versus compatible substances. There is some evidence that the makeup of the MHC can be detected through scent and may influence sexual attraction.

Masculine: Possessing the traits or features that are typically associated with males.

Mate poaching: The attempt to attract and initiate a sexual relationship with an individual who is in already in an established relationship.

Glossary

Mentoplasty: A surgical procedure for reshaping the chin.

Metrosexual: Men who engage in the stereotypically feminine behaviors of caring for their bodies and demonstrating concern about their physical appearance.

Monogamy: The mating strategy of being in a committed relationship with one other individual.

Natural selection: The idea that trait that aids survival will be maintained in a population.

Neuroticism: The Big Five personality dimension that indicates an individual's tendency to feel anxious, depressed, or threatened in social interactions.

Objectification: Judging a person's worth in terms of his or her physical appearance.

Otoplasty: A surgical procedure for reshaping the ears.

Ovulation: The point in the female reproductive cycle when the egg is released from the ovaries and the woman is most likely to conceive.

Parental investment: The amount of time, energy, and resources one invests in raising children.

Personality: The typical psychological qualities and behavioral characteristics of an individual.

Phenotype: The physical expression of the genes into observable traits or behaviors.

Pheromones: A chemical emitted from an individual that may unconsciously influence the physiology or behavior of others.

Physiology: How the cells, organs, and muscles of the body function.

Plastic surgery: Surgery intended to correct the function of a physical feature of the face or body.

Polyandry: The polygamous mating style of a female having multiple husbands.

Polygamy: The mating style of a having multiple spouses.

Polygyny: The polygamous mating style of a male having multiple wives.

Pornography: Sexually explicit material that is made with the intention of sexually arousing the viewer.

Positive reinforcement: Rewarding a behavior to increase the likelihood that the behavior will reoccur in the future.

Posture: How one typically holds one's own body while sitting, standing, or walking.

Primary sex characteristics: The sexual organs responsible for reproduction, including the uterus, vagina, penis, and testes.

Prisoner's dilemma: A game paradigm used to assess the outcomes of selfish or altruistic behavior. For repeated encounters, altruism tends to lead to greater benefits than does selfishness.

Propinquity: Being close to someone in physical space, such as living near another person.

Psychology: The scientific study of the human mind and behavior.

Reproductive fitness: The extent to which one's genes are passed on to the next generation. This may include the number of offspring an individual directly produces or the number of genetically related nieces and nephews who carry the genes.

Rhinoplasty: A surgical procedure for reshaping the nose.

Rhytidectomy: A surgical procedure for lifting and tightening the skin of the face.

Ritual scarring: The process of creating physical scars on the body, typically with the intent of increasing attractiveness.

Sclera: The white part of the eye.

Secondary sex characteristics: The physical differences between men and women that may enhance attractiveness but do not directly influence the reproductive process, such as breasts, voice pitch, body hair, and musculature.

Self-consciousness: Heightened awareness of one's own body, behavior, or appearance.

Self-efficacy: The degree to which individuals feel they can effectively control their own behavior and influence situations.

Self-esteem: Degree of respect and confidence for one's own self-worth and capability.

Self-fulfilling prophecy: The phenomenon where one's expectations positively or negatively influence outcomes. For example, if a woman feels attractive and expects others to perceive her as attractive, she may behave in a more confident and engaging manner, positively influencing their perceptions, and reinforcing her beliefs. Alternatively, if a woman feels unattractive, she may behave in such a way to negatively affect other's perceptions, fulfilling her expectations.

Self-objectification: A common result of objectification when one starts to perceive one's own value in relation to one's appearance rather than cognitive or psychological qualities. Tends to lead to self-consciousness, anxiety, and low self-esteem.

Sex ratio theory: Sex ratio refers to the number of men within a society as compared to the number of women. This ratio is predictive of the sociosexual orientation of the society as well as the typical mating strategy that will emerge within the group.

Sexual dimorphism: The typical differences in physical appearance between the sexes within a species. For example, males tend to be larger than females and have more body and facial hair.

Sexual selection: A specific type of natural selection that predicts that traits that make one more attractive as a mating partner will be maintained in the population.

Glossary

Short-term mating strategy: A mating strategy that lacks the intention to commit to another person and continue the relationship into the future. For example, an individual may engage in a one-night stand without expectations of seeing the other person again.

Shoulder-to-hip ratio (SHR): The ratio of the circumference of the shoulders to the circumference of the hips. Men with high SHR have broad shoulders and narrow hips and are rated as the most physically attractive by women.

Single eyelid (monolid): An eyelid without a crease in the flap of skin over the eye. This type of eyelid is more common in Asian populations and is frequently the target of cosmetic surgery.

Skin bleaching: Lightening the color of one's skin through use of lotions, creams, or bleaches in an attempt to increase the ratings of attractiveness.

Social exchange theory: Suggests that relationships involve exchanges. Successful and satisfying relationships are made up of engaging with others who want what one has to give and who can provide what one needs.

Socioeconomic status (SES): One's comparative level within a social hierarchy, specifically with respect to the level of income, occupation, education, and wealth.

Sociosexual orientation: A rating of the likelihood that an individual will engage in sexual behavior outside of a long-term relationship. Individuals or societies can range from having a restricted sociosexual orientation (monogamous) to an unrestricted sociosexual orientation (promiscuous).

Strategic pluralism theory: The idea that humans have multiple strategies for successful reproduction. The most beneficial strategy changes with respect to environmental pressures. Thus, reproductive strategies may vary between societies or over time.

Stretched lips: A body modification practice in which plugs are inserted into the skin of the bottom lip to gradually increase the size of the lip. The process is predominantly found in South American and African tribes and positively impacts ratings of attractiveness.

Supernormal stimuli: An exaggerated version of a stimulus that evolved to trigger an instinctual response to enhance the response. For example, if men are attracted to a rosy complexion, then rouge may create an exaggerated stimulus to garner even stronger attraction.

Testosterone: The male sex hormone that causes masculinization of male primary and secondary sex characteristics.

Triarchic theory of love: Robert Sternberg's theory that love is composed of three components: passion, intimacy, and commitment. Type of love is determined by which components are present.

Vitruvian Man: Leonardo da Vinci's drawing of the ideal human figure that uses geometry to create ideal proportions based on the descriptions of the ancient Roman architect Vitruvius.

Waist-to-hip ratio (WHR): The comparison of the circumference of the waist compared to the circumference of the hips. A lower WHR would create the hourglass figure considered more attractive for females while a higher WHR would describe a more typically cylindrical male torso. WHR is correlated with attractiveness, health, and reproductive potential for women.

Wingman: A person who lends support when one is attempting to attract a partner.

References and Further Reading

Albino, J. E. N., Lawrence, S. D., & Tedesco, L. A. (1994). Psychological and social effects of orthodontic treatment. *Journal of Behavioral Medicine, 17*(1), 81–98.

American Society of Plastic Surgeons. (2016). *Plastic surgery statistics report. ASPS national clearinghouse of plastic surgery procedural statistics.* Retrieved from https://www.plasticsurgery.org/documents/News/Statistics/2016/plastic-surgery-statistics-full-report-2016.pdf

Anderson, C., John, O. P., Keltner, D., & Kring, A. M. (2001). Who attains social status? Effects of personality and physical attractiveness in social groups. *Journal of Personality and Social Psychology, 81*(1), 116–132.

Antioco, M., Smeesters, D., & Le Boedec, A. (2012). Take your pick: Kate Moss or the girl next door? The effectiveness of cosmetics advertising. *Journal of Advertising Research, 52*(1), 15–30.

Ashikari, M. (2005). Cultivating Japanese whiteness: The "whitening" cosmetics boom and the Japanese identity. *Journal of Material Culture, 10*(1), 73–91.

Baktay-Korsos, G. (1999). The long-hair effect. *Review of Psychology, 6*(1–2), 37–42.

Baudouin, J., & Tiberghien, G. (2004). Symmetry, averageness, and feature size in the facial attractiveness of women. *Acta Pscyhologica, 117*(3), 313–332.

Bereczkei, T., & Mosko, N. (2006). Hair length, facial attractiveness, personality attribution: A multiple fitness model of hairdressing. *Review of Psychology, 13*(1), 35–42.

Berry, D. S., & McArthur, L. Z. (1985). Some components and consequences of a babyface. *Journal of Personality and Social Psychology, 48*(2), 312–323.

Bertamini, M., Byrne, C., & Bennett, K. M. (2013). Attractiveness is influenced by the relationship between postures of the viewer and the viewed person. *i-Perception, 4*(3), 170–179.

Brown, J. D., & L'Engle, K. L. (2009). X-rated: Sexual attitudes and behaviors associated with U.S. early adolescents' exposure to sexually explicit media. *Communication Research, 36*(1), 129–151.

Buss, D. M. (2005). *The handbook of evolutionary psychology.* Hoboken, NJ: Wiley & Sons.

Butler, S. M., Smith, N. K., Collazo, E., Caltabiano, L., & Herbenick, D. (2014). Pubic hair preferences, reasons for removal, and associated genital symptoms: Comparisons between men and women. *The Journal of Sexual Medicine.* Retrieved from http://www.lehmiller.com/

Cheer, L. (2014, August 12). Agonising rites of the crocmen: Boys are cut so that the scars look like crocodile scales in tribal initiation into manhood. *Daily Mail Australia.* Retrieved from http://www.dailymail.co.uk/

Chin, P. K. (2002, January). "If only I could be thin like her, maybe I could be happy like her": The self-implications of associating being thin and attractive with possible life outcomes. *Psychology of Women Quarterly, 27*(3), 209–214.

Clifford, M. M., & Walster, E. (1973). The effect of physical attractiveness on teacher expectations. *Sociology of Education, 46*(2), 248–258.

Commisso, M., & Finkelstein, L. (2012). Physical attractiveness bias in employee termination. *Journal of Applied Social Psychology, 42*(12), 2968–2987.

Diebelius, G. (2016, February 8). How the "perfect" male body has changed over 150 years—from wide waists in the 1870s to the muscular man of today. *Daily Mail.* Retrieved from http://www.dailymail.co.uk/

Dion, K., Berscheid, E., & Waksterm E. (1972). What is beautiful is good. *Journal of Personality and Social Psychology, 4*(3), 285–290.

Dixson, B. J., Dixson, A. F., Morgan, B., & Anderson, M. J. (2007). Human physique and sexual attractiveness: Sexual preferences of men and women in Bakossiland, Cameroon. *Archives of Sexual Behavior, 36*(3), 369–375.

Dixson, B. J., Duncan, M., & Dixson, A. F. (2015). The role of breast size and areolar pigmentation in perceptions of women's sexual attractiveness, reproductive health, sexual maturity, maternal nurturing abilities, and age. *Archives of Sexual Behavior, 44*(6), 1685–1695.

Dixson, B. J., Grimshaw, G. M., Ormsby, D. K., & Dixson, A. F. (2014). Eye-tracking women's preferences for men's somatotypes. *Evolution and Human Behavior, 35*(2), 73–79.

Dixson, B. J., & Rantala, M. J. (2016). The role of facial and body hair distribution in women's judgments of men's sexual attractiveness. *Archives of Sexual Behavior, 45*(4), 877–889.

Dixson, B. J., & Vasey, P. L. (2012). Beards augment perceptions of men's age, social status, and aggressiveness, but not attractiveness. *Behavioral Ecology, 23*(3), 481–490.

Durante, K. M., Griskevicius, V., Simpson, J. A., Canfu, S. M., & Li, N. P. (2012). Ovulation leads women to perceive sexy cads as good dads. *Journal of Personality and Social Psychology, 103*(2), 292–305.

Edmonds, A. (2012). Body image in non-Western societies. *Encyclopedia of body image and human appearance, Volume 1.* Waltham, MA: Elsevier Inc.

Efran, M. G. (1974). The effect of physical appearance on the judgment of guilt, interpersonal attraction, and severity of recommended punishment in simulated jury task. *Journal of Research in Personality, 8*(1), 45–54.

Engel, P. (2014, May 25). MAP: Divorce rates around the world. *Business Insider.* Retrieved from http://www.businessinsider.com/

Evans, P. C., & McConnell, A. R. (2003). Do racial minorities respond in the same way to mainstream beauty standards? Social comparison processes in Asian, black, and white women. *Self and Identity, 2*(2), 153–167.

Feng, C. (2002). Looking good: The psychology and biology of beauty. *Journal of Young Investigators, 6*(6), 1–4.

Filiault, S. M. (2007). Measuring up in the bedroom: Muscle, thinness, and men's sex lives. *International Journal of Men's Health, 6*(2), 127–142.

Fink, B., Grammer, K, & Thornhill, R. (2001). Human (*Homo sapiens*) facial attractiveness in relation to skin texture and color. *Journal of Comparative Psychology, 115*(1), 92–99.

Fink, B., & Neave, N. (2005). The biology of facial beauty. *International Journal of Cosmetic Science, 27*(6), 317–325.

Foote, K. (2014, Mar 11). Ethnic travel: Should you visit Thailand's long neck women villages? *Epicure & Culture.* Retrieved from http://epicureandculture.com/

Foreman, A. (2015, February). Why footbinding persisted in China for a millennium: Despite the pain, millions of Chinese women stood firm in their devotion to the tradition. *Smithsonian Magazine.* Retrieved from https://www.smithsonianmag.com/

Fredrickson, B. L., Roberts, T. A., Noll, S. M., Quinn, D. M., & Twenge, J. M. (1998). That swimsuit becomes you: Sex differences in self-objectification, restrained eating, and math performance. *Journal of Personality and Social Psychology, 75*(1), 269–284.

Friedman, V. (2017, May 8). A new age in French—modeling. *The New York Times.* Retrieved from https://www.nytimes.com

Frith, K., Shaw, P., & Cheng, H. (2005). The construction of beauty: A cross-cultural analysis of women's magazine advertising. *Journal of Communication, 55*(1), 56–70.

Furnham, A. (2015, April 2). The attractiveness of personality traits: Is an attractive personality more important than an attractive body? *Psychology Today.* Retrieved from https://www.psychologytoday.com

Furnham, A., & Swami, V. (2007). Perception of female buttocks and breast size in profile. *Social Behavior and Personality, 35*(1), 1–8.

Gangestad, S. W., & Thornhill, R. (1998). Menstrual cycle variation in women's preference for the scent of symmetrical men. *Proceedings of the Royal Society of London Series B, 265*(1399), 927–933.

Garver-Apgar, C. E., Gangestad, S. W., Thornhill, R., Miller, R. D., & Olp, J. J. (2006). Major histocompatibility complex alleles, sexual responsivity, and unfaithfulness in romantic couples. *Psychological Science, 17*(10), 830–835.

Geldart, S. (2008). Tall and good-looking? The relationship between raters' height and perceptions of attractiveness. *Journal of Individual Differences, 29*(3), 148–156.

Gildersleeve, K. A., Haselton, M. G., Larson, C. M., & Pillsworth, E. G. (2012). Body odor attractiveness as a cue of impending ovulation in women: Evidence from a study using hormone-confirmed ovulation. *Hormones and Behavior, 61*(2), 157–166.

Goldey, K. L., Avery, L. R., & van Anders, S. M. (2014). Sexual fantasies and gender/sex: A multimethod approach with quantitative content analysis and hormonal responses. *Journal of Sex Research, 51*(8), 917–931.

Goldman, B. (2013, December 23). In men, high testosterone can mean weakened immune system response, study finds. *Stanford Medicine*. Retrieved from: https://med.stanford.edu/.

Greitemeyer, T. (2005). Receptivity to sexual offers as a function of sex, socioeconomic status, physical attractiveness, and intimacy of the offer. *Personal Relationships, 12*(3), 373–386.

Grosofsky, A., Adkins, S., Bastholm, R., Meyer, L, Krueger, L., Meyer, J., & Torma, P. (2003). Tooth color: Effects on judgments of attractiveness and age. *Perceptual and Motor Skills, 96*(1), 43–48.

Guéguen, N. (2013). Effects of a tattoo on men's behavior and attitudes towards women: An experimental field study. *Archives of Sexual Behavior, 42*(8), 1517–1524.

Guéguen, N. (2008). The effects of women's cosmetics on men's approach: An evaluation in a bar. *North American Journal of Psychology, 10*(1), 221–228.

Guizzo, F., & Gadinu, M. (2017). Effects of objectifying gaze on female cognitive performance: The role of flow experience and internalization of beauty ideals. *The British Journal of Social Psychology, 56*(2), 281–292.

Gupta, N. D., Etcoff, N. L., Jaeger, M. M. (2016). Beauty in the mind: The effects of physical attractiveness on psychological well-being and distress. *Journal of Happiness Studies, 17*(3), 1313–1325.

Guynup, S. (2004, July 28). Scarification: Ancient body art leaving new marks. *National Geographic*. Retrieved from http://news.nationalgeographic.com/

Hald, G. M. (2006). Gender differences in pornography consumption among young heterosexual Danish adults. *Archives of Sexual Behavior, 35*(5), 577–585.

Halliwell, E., & Dittmar, H. (2004). Does size matter? The impact of model's body size on women's body-focused anxiety and advertising effectiveness. *Journal of Social and Clinical Psychology, 23*(1), 104–122.

Harter, P. (2004, January 26). Mauritania's "wife-fattening" farm. *BBC News*. Retrieved from http://news.bbc.co.uk

Hartmann, A. S., Thomas, J. J., Greenberg, J. L., Elliott, C. M., Matheny, N. L., & Wilhelm, S. (2015). Anorexia nervosa and body dysmorphic disorder: A comparison of body image concerns and explicit and implicit attractiveness beliefs. *Body Image, 14*, 77–84.

Haselton, M. G., Mortezaie, M., Pillsworth, E. G., Bleske-Rechek, A., & Frederick D. A. (2007). Ovulatory shifts in human female ornamentation: Near ovulation, women dress to impress. *Hormones and Behavior, 51*(1), 40–45.

Haworth, A. (2001, Jul 20). Forced to be fat: Forcefeeding in Mauritania—West Africa fat camp. *Marie Claire*. Retrieved from http://www.marieclaire.com/

Heinberg, L. J., & Thompson, J. K. (1995). Body image and televised images of thinness and attractiveness: A controlled laboratory investigation. *Journal of Social and Clinical Psychology, 14*(4), 325–338.

Henig, R. M. (1996 May/June). The price of perfection. *Civilization, 3*(3), 56–61.

Holman, A. (2011). Psychology of beauty: An overview of the contemporary research lines. *Social Psychology, 28*, 81–94.

Hosoda, M., Stone-Romero, E. F., & Coats, G. (2003). The effects of physical attractiveness on job-related outcomes: A meta-analysis of experimental studies. *Personnel Psychology, 56*(2), 431–462.

Husain, W., & Qureshi, Z. (2016). Preferences in marital sexual practices and the role of pornography. *Sexologies: European Journal of Sexology and Sexual Health/Revue Européenne De Sexologie Et De Santé Sexuelle, 25*(2), 35–41. doi:10.1016/j.sexol.2016.01.005

Hwang, H. S., & Spiegel, J. H. (2014). The effect of "single" vs "double" eyelids on the perceived attractiveness of Chinese women. *Aesthetic Surgery Journal, 34*(3), 374–382.

Johnston, V. S., Hagel, R., Franklin, M., Fink, B., & Grammer, K. (2001). Male facial attractiveness evidence for hormone-mediated adaptive design. *Evolution and Human Behavior, 22*(4), 251–267.

Jones, A. L., Russell, R., & Ward, R. (2015). Cosmetics alter biologically-based factors of beauty: Evidence from facial contrast. *Evolutionary Psychology, 13*(1), 210–229.

Jones, B. A., & Griffiths, K. M. (2015). Self-objectification and depression: An integrative systematic review. *Journal of Affective Disorders, 171*, 22–32.

Jones, D. (1995). Sexual selection, physical attractiveness, and facial neoteny. *Current Anthropology, 36*(5), 723–748.

Kalodner, C. R. (1997). Media influences on male and female non-eating-disordered college students: A significant issue. *Eating Disorders, 5*(1), 47–57.

Keating, C. F. (1985). Gender and the physiognomy of dominance and attractiveness. *Social Psychology Quarterly, 48*(1), 61–70.

Kohl, J. V., Atzmueller, M., Fink, B., & Grammer, K. (2007). Human pheromones: Integrating neuroendocrinology and ethology. *Homeostasis and Disease, 49*(3/4), 123–135.

Kościński, K. (2012). Hand attractiveness—Its determinants and associations with facial attractiveness. *Behavioral Ecology, 23*(2), 334–342.

Kremer, W. (2013, January 25). Why did men stop wearing high heels? *BBC World Service*. Retrieved from http://www.bbc.com/

Kubota, T. (2017). The connection between scent and sexual attraction. *Men's Journal*. Retrieved from http://www.mensjournal.com/

Laeng, B., Mathisen, R., & Johnsen, J. (2007). Why do blue-eyed men prefer women with the same eye color? *Behavioral Ecology and Sociobiology, 61*, 371–384.

Langlois, J. H., Kalakanis, L., Rubenstein, A. J., Larson, A., Hallam, M., & Smoot, M. (2000). Maxims or myths of beauty? A meta-analytic and theoretical review. *Psychological Bulletin, 126*(3), 390–423.

LaTosky, S. L. (2004). Reflections on the lip plates of Mursi women as a source of stigma and self-esteem. In Ivo Strecker & Jean Lydall (Eds.), *The perils of *face: Essays on cultural contact, respect, and self-esteem in southern Ethiopia* (382–397). Munster: LIT Verlag.

Little, A. C., & Hancock, P. J. B. (2002). The role of masculinity and distinctiveness in judgments of human male facial attractiveness. *British Journal of Psychology, 93*(4), 451–464.

Little, A. C., Jones, B., C., & DeBruine, L. M. (2011, June 12). Facial attractiveness: Evolutionary based research. *Philosophical Transactions of the Royal Society B: Biological Sciences, 366*(1571), 1638–1659.

Lübke, K. T., & Pause, B. M. (2015). Always follow your nose: The functional significance of social chemosignals in human reproduction and survival. *Hormones and Behavior, 68,* 134–144.

Lundborg, P., Nystedt, P., & Lindgren, B. (2007). Getting ready for the marriage market? The association between divorce risks and investments in attractive body mass among married Europeans. *Journal of Biosocial Science, 39*(4), 531–544.

Luo, W. (2013). Aching for the altered body: Beauty economy and Chinese women's consumption of cosmetic surgery. *Women's Studies International Forum, 38,* 1–10.

Lydon, J. E., Meana, M., Sepinwall, D., Richards, N., & Mayman, S. (1999). The commitment calibration hypothesis: When do people devalue attractive alternatives? *Personality and Social Psychology Bulletin, 25*(2), 152–161.

Ma, F., Xu, F, & Luo, X. (2015). Children's and adults' judgments of facial trustworthiness: The relationship to facial attractiveness. *Perceptual and Motor Skills, 121*(1), 179–198.

Macia, E., Duboz, P., & Cheve, D. (2015). The paradox of impossible beauty: Body changes and beauty practices in older women. *Journal of Women and Aging, 27*(2), 174–187.

Magee, L. (2012). Cosmetic surgical and non-surgical procedures for the face. *Encyclopedia of body image and human appearance, Volume 1.* Waltham, MA: Elsevier Inc.

Ma-Kellams, C., Wang, M. C., & Cardiel, H. (2017). Attractiveness and relationship longevity: Beauty is not what it is cracked up to be. *Personal Relationships, 24*(1), 146–161.

Marino Carper, T. L., Negy, C., & Tantleff-Dunn, S. (2010). Relations among media influence, body image, eating concerns, and sexual orientation in men: A preliminary investigation. *Body Image, 7*(4), 301–309.

Mathes, E. W., & Kozak, G. (2008). The exchange of physical attractiveness for resource potential and commitment. *Journal of Evolutionary Psychology, 6*(1), 43–56.

Mauritania struggles with love of fat women: Government trying to change desert culture that force-feeds girls (2007, April 16). *NBC News.* Retrieved from http://www.nbcnews.com/

Mazur, A. (2986). U.S. trends in feminine beauty and overadaptation. *The Journal of Sex Research, 22*(3), 281–303.

McGovern, R. J., Neale, M. C., & Kendler, K. S. (1996). The independence of physical attractiveness and symptoms of depression in a female twin population. *The Journal of Psychology, 130*(2), 209–219.

Mckay, A. (2014). *Western beauty pressures and their impact on diverse women*. Unpublished honors thesis, Brock University, Ontario, Canada. Retrieved from https://brocku.ca/webfm_send/32237

Meerdink, J. E., Garbin, C. P., & Leger, D. W. (1990). Cross-gender perceptions of facial attributes and their relation to attractiveness: Do we see them differently than they see us? *Perception & Psychophysics, 48*(3), 227–233.

Miller, S. L., & Maner, J. K. (2010). Evolution and relationship maintenance: Fertility cues lead committed men to devalue relationship alternatives. *Journal of Experimental Social Psychology, 46*(6), 1081–1084.

Morris, P. H., White, J., Morrison, E. R., & Fisher, K. (2013). High heels as supernormal stimuli: How wearing high heels affects judgements of female attractiveness. *Evolution and Behavior, 34*(3), 176–181.

Morrow, P. (1990). Physical attractiveness and selection decision-making. *Journal of Management, 16*(1), 45–60.

Mosher, D. L. (1988). Pornography defined: Sexual involvement theory, narrative context, and goodness-of-fit. *Journal of Psychology & Human Sexuality, 1*(1), 67–85.

Moyo, M. (2015, July 13). The Karo tribe: Ethiopia's indigenous group that excels in body painting and scarification. *Nomad*. Retrieved from http://www.nomadafricamag.com/

Mulford, M., Orbell, J., Shatto, C., and Stockard, J. (1998). Physical attractiveness, opportunity, and success in everyday exchange. *American Journal of Sociology, 103*(6), 1565–1592.

Nadler, A., & Dotan, I. (1992). Commitment and rival attractiveness: Their effects on male and female reactions to jealousy-arousing situations. *Sex Roles, 26*(7–8), 293–310.

Neva. (2013, May 7). Tribe: The Mangbetu—The head elongation fashionistas of central Africa. *Afritorial*. Retrieved from http://afritorial.com/

Nicholson, J. (2011, May 24). Is your personality making you more or less physically attractive? Does your personality influence how physically attractive you are? *Psychology Today*. Retrieved from https://www.psychologytoday.com

Noles, S. W., Cash, T. F., Winstead, B. A. (1985). Body image, physical attractiveness, and depression. *Journal of Consulting and Clinical Psychology, 53*(1), 88–94.

Noor, F., & Evans, D. C. (2003). The effect of facial symmetry on perceptions of personality and attractiveness. *Journal of Research in Personality, 37*(4), 339–347.

Osborn, D. R., (1996). Beauty is as beauty does?: Makeup and posture effects on physical attractiveness judgments. *Journal of Applied Social Psychology, 26*(1), 31–51.

Pazda, A. D., Elliot, A. J., & Greitemeyer, T. (2012). The color of sexuality: Female red displays are used and perceived as a sexual signal. *Advances in Psychology Research, 89*, 145–155.

Peters, M., Rhodes, G., & Simmons, L. W. (2008). Does attractiveness in men provide clues to semen quality? *Journal of Evolutionary Biology, 21*, 572–579.

Pinhas, L., Toner, B. B., Ali, A., Garfinkel, P. E., & Stuckless, N. (1999). The effects of the ideal of female beauty on mood and body satisfaction. *International Journal of Eating Disorders, 25*(2), 223–226.

Poloskov, E., & Tracey, T. G. (2013). Internalization of U.S. female beauty standards as a mediator of the relationship between Mexican American women's acculturation and body dissatisfaction. *Body Image, 10*(4), 501–508.

Pornography statistics: 2015 report. (2015). Retrieved from http://www.covenant eyes.com/pornstats/

Presser, B. (2014, August 11). Facial tattoos: The tribal female rite in Papua New Guinea. *The Daily Beast.* Retrieved from https://www.thedailybeast.com/

Prokop, P., & Fedor, P. (2011). Physical attractiveness influences reproductive success of modern men. *Journal of Ethology, 29*(3), 453–458.

Ramirez, A. (1990, August 12). All about deodorants; the success of sweet smell. *New York Times.* Retrieved from http://www.nytimes.com/

Rantala, M. J., Eriksson, C. J. P., Vainikka, A., & Kortet, R. (2006). Male steroid hormones and female preference for male body odor. *Evolution and Human Behavior, 27*(4), 259–269.

Remoff, H. T. (1984). *Sexual choice: A woman's decision.* New York: Dutton-Lewis Publishing.

Rhodes, G. (2006). The evolutionary psychology of facial beauty. *Annual Review of Psychology, 57,* 199–226.

Ricciardelli, L. A., & Williams, R. J. (2012). Beauty over the centuries—Male. *Encyclopedia of body image and human appearance, Volume 1.* Waltham, MA: Elsevier Inc.

Rich, M. K., & Cash, R. F. (1993). The American image of beauty: Media representations of hair color for four decades. *Sex Roles, 29*(1/2), 113–124.

Roberts, S. C., Little, A. C., Lyndon, A., Roberts, J., Havlicek, J., & Wright, R. L. (2009). Manipulation of body odour alters men's self-confidence and judgements of their visual attractiveness by women. *International Journal of Cosmetic Science, 31*(1), 47–54.

Rollero, C. (2015). "I know you are not real": Salience of photo retouching reduces the negative effects of media exposure via internalization. *Studia Psychologica, 57*(3), 195–202.

Ryan, R. (2014, August 11). Crocodile scarification is an ancient initiation practised by the Chambri tribe of Papua New Guinea. *NT News.* Retrieved from http://www.ntnews.com.au/

Saegusa, C., & Watanabe, K. (2016). Judgments of facial attractiveness as a combination of facial parts information over time: Social and aesthetic factors. *Journal of Experimental Psychology: Human Perception and Performance, 42*(2), 173–179.

Salusso-Deonier, C. J., Markee, N. L., & Pedersen, E. L. (1993). Gender differences in the evaluation of physical attractiveness ideals for male and female body builds. *Perceptual and Motor Skills, 76*(3), 1155–1167.

Sammaknejad, N. (2013). *Facial attractiveness: The role of iris size, pupil size, and scleral color* (Doctoral dissertation). Ann Arbor, MI: UMI Dissertation Publishing.

Schreiber, K. (2015). Promoting a thin and ultra-athletic physique has unforeseen consequences. *Psychology Today.* Retrieved from https://www.psychologyto day.com/articles/201509/mindyourbodybodyconscious

Scodel, A. (1957). Heterosexual somatic preference and fantasy dependency. *Journal of Consulting Psychology, 21*(5), 371–374.

Scott, I. M. L., Pound, N., Stephen, I. D., Clark, A. P., & Penton-Voak, I. S. (2010). Does masculinity matter? The contribution of masculine face shape to male attractiveness in humans. *Public Library of Science, 5*(10), 1–10.

Sentilles, R. M., & Callahan, K. (2012). Beauty over the centuries—Female. *Encyclopedia of body image and human appearance, Volume 1.* Waltham, MA: Elsevier Inc.

Shackelford, T. K., & Larsen R. J. (1997). Facial asymmetry as an indicator of psychological, emotional, and physiological distress. *Journal of Personality and Social Psychology, 72*(2), 456–466.

Shahani, C, Dipboye, R. L., & Gehrlein, T. M. (1993). Attractiveness bias in the interview: Exploring the boundaries of an effect. *Basic and Applied Social Psychology, 14*(3), 317–328.

Shapouri, B. (2016, May 26). The way we buy beauty now: How millennial skepticism is revolutionizing the beauty industry, one purchase at a time. *Racked.* Retrieved from http://www.racked.com/2016/5/26/11674106/buyingbeauty sephoradepartment

Singh, D., Dixson, B. J., Jessop, T. S., Morgan, B., & Dixson, A. F. (2010). Cross-cultural consensus for waist-hip ratio and women's attractiveness. *Evolution and Human Behavior, 31*(3), 176–181.

Singh, D., & Singh, D. (2006). Role of body fat and body shape on judgment of female health and attractiveness: An evolutionary perspective. *Psychological Topics, 15*(2), 331–350.

Sobania, N. W. (2003). *Culture and customs of Kenya.* Westport, CT: Greenwood Publishing Group.

Sorokowski, P., Sorokowska, A., & Mberira, M. (2012). Are preferences for legs length universal? Data from a semi-nomadic Himba population from Namibia. *The Journal of Social Psychology, 152*(3), 370–378.

Sorokowski, P., Szmajke, A., Sorokowska, A., Cunen, M.B., Fabrykant, M., Zarafshani, K., . . . Fang, T. (2011). Attractiveness of leg length: Report from 27 nations. *Journal of Cross-Cultural Psychology, 42*(1), 131–139.

Staat, J. C. (1977). *Size of nose and mouth as components of facial beauty* (Doctoral dissertation). Ann Arbor, MI: Xerox University Microfilms.

Staley, C., & Prause, N. (2013). Erotica viewing effects on intimate relationships and self/partner evaluations. *Archives of Sexual Behavior, 42*(4), 615–624.

Stephen, I. D., & Mckeegan, A. M. (2010). Lip colour affects perceived sex typicality and attractiveness of human faces. *Perception, 39*(8), 1104–1110.

Sugiyama, L. (2005). Physical attractiveness in adaptationist perspective. In D. M. Buss (Ed.), *The handbook of evolutionary psychology* (pp. 292–343). Hoboken, NJ: Wiley & Sons.

Sun, C., Bridges, A., Johnson, J. A., & Ezzell, M. B. (2016). Pornography and the male sexual script: An analysis of consumption and sexual relations. *Archives of Sexual Behavior, 45*(4), 983–994.

Švegar, D. (2016). What does facial symmetry reveal about health and personality? *Polish Psychological Bulletin, 47*(3), 356–365.

Svoboda, E. (2008). Scents and sensibility. *Psychology Today*. Retrieved from https://www.psychologytoday.com

Swami, V., Furnham, A., & Joshi, K. (2008). The influence of skin tone, hair length, and hair colour on ratings of women's physical attractiveness, health and fertility. *Scandinavian Journal of Psychology, 49*(5), 429–437.

Sypeck, M. F., Gray, J. J., & Ahrens, A. H. (2004). No longer just a pretty face: Fashion magazines' depictions of ideal female beauty from 1959–1999. *International Journal of Eating Disorders, 36*(3), 342–347.

Tartaglia, S., & Rollero, C. (2015). The effects of attractiveness and status on personality evaluation. *Europe's Journal of Psychology, 11*(4), 677–690.

Thorndike, E. L. (1920). The constant error in psychological ratings. *Journal of Applied Psychology, 4*(1), 25–29.

Thornhill, R., Gangestad, S. W., Miller, R., Scheyd, G., McCollough, J. K., & Franklin, M. (2003). Major histocompatibility complex genes, symmetry, and body scent attractiveness in men and women. *Behavioral Ecology, 14*(5), 668–678.

Tiggemann, M., & Hodgson, S. (2008). The hairlessness norm extended: Reasons for and predictors of women's body hair removal at different body sites. *Sex Roles, 59*(11–12), 889–897.

Tiggemann, M., & Lewis, C. (2004). Attitudes toward women's body hair: Relationship with disgust sensitivity. *Psychology of Women Quarterly, 28*(4), 381–387.

Taormina, R. J., & Ho, I. M. (2012). Intimate relationships in China: Predictors across genders for dating, engaged, and married individuals. *Journal of Relationships Research, 3*, 24–43.

Taylor, L. D., & Fortaleza, J. (2016). Media violence and male body image. *Psychology of Men and Masculinity, 17*(4), 380–384.

Træen, B., Štulhofer, A., & Carvalheira, A. (2013). The associations among satisfaction with the division of housework, partner's perceived attractiveness, emotional intimacy, and sexual satisfaction in a sample of married or cohabiting Norwegian middle-class men. *Sexual and Relationship Therapy, 28*(3), 215–229.

Trekels, J., & Eggermont, S. (2017). Linking magazine exposure to social appearance anxiety: The role of appearance norms in early adolescence. *Journal of Research on Adolescence, 27*(4), 736–751.

Vandenbosch, L., & Eggermont, S. (2012). Understanding sexual objectification: A comprehensive approach toward media exposure and girls' internalization of beauty ideals, self-objectification, and body surveillance. *Journal of Communication, 62*(5), 869–887.

Van der Geld, P., Oosterveld, P., Van Heck, G., & Kuijpers-Jagtman, A. M. (2007). Smile attractiveness: Self-perception and influence on personality. *The Angle Orthodontist, 77*(5), 759–765.

Vashi, N. A. (2015). *Beauty and body dysmorphic disorder*. Switzerland: Springer International Publishing.

Wack, E. R., & Tantleff-Dunn, S. (2008). Cyber sexy: Electronic game play and perceptions of attractiveness among college-aged men. *Body Image, 5*(4), 365–374.

Wade, T. J. (2010). The relationships between symmetry and attractiveness and mating relevant decisions and behavior: A review. *Symmetry, 2*, 1081–1098.

Wang, A. (2009). *Physical attractiveness and its effects on social treatment and inequality.* Available at SSRN: http://ssrn.com/abstract=1518099

Wegenstein, B. (2012). *The cosmetic gaze: Body modification and the construction of beauty.* Cambridge, MA: MIT Press.

Wiederman, M. W., & Dubois, S. L. (1998). Evolution and sex differences in preferences for short-term mates: Results from a policy capturing study. *Evolution and Human Behavior, 19*(3), 153–170.

Weinberg, M. S., Williams, C. J., Kleiner, S., & Irizarry, Y. (2010). Pornography, normalization, and empowerment. *Archives of Sexual Behavior, 39*(6), 1389–1401.

Wogalter, M. S., & Hosie, J. A. (1991). Effects of cranial and facial hair on perceptions of age and person. *The Journal of Social Psychology, 131*(4), 589–591.

Wohlrab, S., Find, B., Kappeler, P. M., & Brewer, G. (2009). Perception of human body modification. *Personality and Individual Differences, 46*, 202–206.

Yu, S. Y., Park, E., & Sung, M. (2015). Cosmetics advertisements in women's magazines: A cross-cultural analysis of China and Korea. *Social Behavior and Personality, 43*(4), 685–704.

Zebrowitz, L. A., & Montepare, J. M. (2008). Social psychological face perception: Why appearance matters. *Social Personal Psychological Compass, 2*(3), 1497–1512.

Zelazniewicz, A. M., & Pawlowski, B. (2011). Female breast size attractiveness for men as a function of sociosexual orientation (restricted vs. unrestricted). *Archives of Sexual Behavior, 40*(6), 1129–1135.

Index

Abdominal muscles, 54, 218
Aboriginal culture of Australia, 177
Academic performance, 38, 40, 48
Adjustment, 49–50
Advertisements: changes in advertising, 82; cultural comparisons, 28, 142–43, 151–52, 180; dental advertisements, 202; impacts on psychological health, 160–64; Photoshop, 205; sexual appeal in advertising, 95–96
African, 5, 7, 180, 211; African American, 12–13, 154–55; black women, 152–54
Age: and attractiveness, 3, 5, 10–12, 24, 26, 38, 86–88, 165; and beards, 172; and body satisfaction, 142, 147, 157, 163; and breasts, 215; and chins, 188; and dieting, 142; and disorders, 32; and ears, 195; and elongated necks, 212; and eyebrows, 61, 175–76; and eyes, 193–94; and feet, 220; and femininity, 62; and hair, 64–65, 165–68; and hair loss, 173; and lips, 201; and marriage, 130; and media, 92, 151; and obesity, 206–7; of perceiver/rater, 27, 37, 46, 73, 176; and pornography use, 137–38; and reproduction, 98–99, 105; and sex, 137–38; and skin, 178–79, 181; and teeth, 202–3
Aggressiveness, 46, 171, 208
Agreeableness, 4, 18–23, 124–25
Altruistic behavior, 44
Ambition, 4, 18, 103
Ancient Greeks, 13, 78–79, 91
Ancient Romans, 13, 78–79, 91, 93
Androgynous, 13, 225
Anemia, 33
Anger, 121
Anorexia, 13, 82, 159–60
Anthropology, 3, 12, 18, 29, 203, 212, 221
Anxiety, 23, 70, 125, 141, 147–49, 154, 156, 159–60, 162–64
Arranged marriage, 26, 101–2
Asian: beauty ideals, 143, 150; cosmetic surgery, 143, 154, 190–92; media, 28, 144, 150–52; self-esteem, 153–54; trustworthiness, 46
Assertiveness, 35, 40, 47, 60, 87, 201

Asymmetry, 5, 7–8, 22–23, 66, 170, 192
Attractiveness: age, 10–12, 86–88; anxiety, 163–64; averageness, 5–8, 27, 185; baldness, 66, 172–73; bathing, 78–79; BMI, 12–13, 204–8; body image, 156–59; body shape, 13–15, 104–6; breasts and buttocks, 215–17; commitment, 120–21, 127–29; consequences, 88–89, 94–95; context, 106; cosmetic surgery, 62–64; cosmetics, 56–62, 76–78; cultural similarities, 100–111; depression, 161–63; divorce, 129–33; ears, 195–97; eating disorders, 159–61; environmental influences, 111–15; eyebrows, 175–76; eyes, 189–94; facial features, 5–6, 184–89; fashion, 73–75, 85–86; feet, 219–21; friendships, 106–7; gender, 101–6; gym, 71–72; hair, 64–67, 93–94, 165–72; hair removal, 173–74; hands, 70–71, 203–4; health and reproduction, 32–33; height, 211–15; hormones, 221–31; ideals, 82–95; intimacy, 123–27; jealousy, 120–23; lips, 197–200; long-term relationships, 110–11; LTR, 210–11; media, 97–98, 142–56; muscularity, 217–18; nose, 194–95; parental investment, 100–101; penis, 219; personality, 17–24, 231–32; perspectives, 26–29; pheromones, 16–17, 68–70, 225–30; pornography, 136–39; propinquity, 24–26; psychological effects, 141–64; resources, 102–4; scars, 177; scent, 16–17, 68–70, 225–30; sex ratio, 112; sexual fantasies, 133–35; sexual satisfaction, 126–27; sexual selection, 100–101; short-term relationships, 107–10, 117–20; SHR, 13–15, 210; similarity, 24–26; skin, 176–83; smile, 67–68; social benefits, 33–50; social exchange theory, 113–15; symmetry, 5–9, 17–23; tattoos, 75–76, 78; teeth, 200–203; vocal quality, 17, 230–31; WHR, 13–15, 208–10; wingman, 72–73. *See also* Beauty
Australia, 7–8, 94, 173, 177, 186, 218
Averageness, 5–7, 27, 185, 194, 204, 211, 221

Babies, 3, 5, 186, 193
Baby face, 43, 188
Baldness, 66, 172–73
Bathing, 68–69, 79
Bathing suits, 85; swimwear, 85
Beards, 92–93, 170–73. *See also* Facial hair
Beauty: age, 11, 86–88; beauty products, 53, 87; benefits in adulthood, 41–42; benefits in childhood, 36–38; benefits in college, 39–41; body dissatisfaction, 154; body image, 156–59; BMI, 12, 106, 145; buttocks, 217; celebrities, 28; changes over time for men, 89–95; changes over time for women, 82–89; competence, 170; consequences for men, 94–95; consequences for women, 88–89; cooperation and competition, 44–46; cosmetic surgery, 63; cosmetics, 56–61, 143, 155; cross-cultural, 28, 150–56; defining beauty, 3–29; depression, 161–63; ears, 196; eating disorders, 159–61; elongated skulls, 186–87; eyes, 190, 193; facial beauty, 36, 168; facial expressions, 55; familiarity, 25; fashion, 85–86; fashion magazines, 145–46; feet, 220; fertility, 4, 32–33, 53, 114; genetic quality, 26; Girl Scouts, 83; hair, 168; halo effect, 33–36; historic methods, 76–78; lips, 198–99; media, 141–48, 152; music videos, 149; natural beauty, 54–56, 79; neck, 212; nose, 194; occupational success,

46–48; pageants, 88; perspectives, 26–29; pheromones, 16, 65, 69, 222, 228–29; Photoshop, 149–50; physical health, 5, 8, 31–33, 49–50; psychological health, 141–64; romantic relationships, 42–44; scars, 177; scent, 17; sexual dimorphism, 61; skin, 81; skin bleaching, 180; status, 82, 177, 180, 186–87, 196, 198; symmetry, 27; tattoos, 182; trends, 81–96; types, 150–51; weight, 207; WHR, 209. *See also* Attractiveness

Beauty standards, 63, 86, 94, 141, 143, 151–53, 157, 159

Behavioral features, 105

Benefits of beauty: in adulthood, 41–42; in childhood, 36–38; in college, 39–41; for occupational success, 46–48; for romantic relationships, 42–44

Big-Five Personality Traits: agreeableness, 4, 18–23, 124–25; conscientiousness, 18–23; extraversion, 18–23, 35, 40, 124; neuroticism, 18–23, 40, 124, 201; openness to new experience, 18–21

Biological drive, 4, 81

Biological perspective, 27, 31

Blepharoplasty, 62, 190

Body consciousness, 96, 207

Body fat, 15, 91, 104, 204, 208–9, 218

Body hair: attractiveness ratings, 92; health, 174; males, 170–71; placement, 66, 165; removal, 65–67, 94, 173–74

Body image: anxiety, 163; cosmetic surgery, 95; cosmetics, 63; cultural differences and pressures, 154, 158, 204; depression, 162; eating disorders, 160–61; effect of advertisements, 96; Girl Scouts, 83; media, 141–42, 145–48, 157, 159, 164; men, 217–18; objectification, 156; Photoshop, 150; pornography, 136–38

Body mass index (BMI), 12–15, 204–10; attractiveness, 11, 54, 92; cultural differences, 105–6; divorce, 132; health, 32–33, 50

Body modification, 147, 195, 197

Body odor, 16, 28, 69–70, 222, 228–30

Body scent: attractiveness, 3, 16, 229; confidence, 228; deodorants, 68–69; inguinal crease, 218; perfume and cologne, 69–70, 228; symmetry, 8. *See also* Pheromones

Body shape: attractiveness, 53, 56, 63, 92, 165; BMI, 13, 204–10; exercise, 54; fashion, 74; health, 72, 95, 232; ideals, 82, 92; objective, 3; SHR, 13–15; WHR, 13–15, 21, 32

Body symmetry, 5, 8, 32, 90–91, 96, 104, 119, 177

Breasts: age, 11, 215; attractiveness, 216; augmentation/implants, 62–63, 85, 88, 150, 217; body scent, 16; cancer, 32; changes in ideals, 82–86; fashion, 64, 83–86; health, 8, 215; media, 148; size and shape, 215–17; symmetry, 8, 215; tipping behavior, 60

Bulimia, 13, 159

Burn the bra, 83

Buss, David, 10, 101–4, 108–10, 113, 131–32, 230

Buttocks, 74, 183, 209, 215–17

Camaraderie, 21

Cancer, 32–33, 150, 180, 209

Cardiovascular health, 71–72, 208

Caucasian: BMI, 12–13; double eye-lid, 63; eating disorders, 13; models, 64, 143–44, 150–56; ratings of attractiveness, 5, 179, 222

Chastity, 105

Cheekbones, 9, 46, 56, 61, 104, 184–85, 188, 225

Chin, 5, 42–43, 46, 55, 62, 184, 188, 193, 222
China: arranged marriages, 101; cosmetic surgery, 63, 190; foot binding, 220–21; interdependence, 35; modeling, 150; preferences, 171
Closed individuals, 19
Clothing color, 74–75
Cognitive script, 137
Cologne, 57, 68–70, 79, 228–29
Competition, 98, 115, 198
Competence, 37, 41, 47, 49–50, 60, 66, 160, 170
Complexion, 53–54, 57, 77, 82, 154, 176–81
Confederate, 122, 222, 230
Confidence: achievement, 34; advertisements/media, 143, 145, 146, 157, 202; attractiveness, 40–43, 51, 59, 141, 162, 168; beards, 93, 172; body odor, 228–29; cosmetic surgery, 62, 191–92; cosmetics, 60–61, 79; height, 214; models, 83; nose, 195; paternity, 189; penis size, 219; personality, 18; Photoshop, 150; pornography, 136, 139; posture, 55; support, 126; teeth, 201, 203
Conflict resolution, 123
Conscientiousness, 18–23
Consequences of beauty ideals for men, 94
Consequences of beauty ideals for women, 88
Contraceptives, 16, 137, 224, 227
Cooperation, 21, 35, 41, 44–45
Cooperative alliances, 106
Corsets, 73, 82–86, 93, 221
Cosmetics, 56–62, 79, 85; age, 87, 175–76; cultural differences, 143, 154, 180; history of cosmetics, 76–78; lips, 200; perfume and cologne, 69; skin texture, 181
Cosmetic surgery, 62–64, 78, 85, 87–88, 115; Asian populations, 143, 154, 190; blepharoplasty, 62, 190; breast augmentation, 62–63, 217; impact of Photoshop, 149–50; labiaplasty, 63; liposuction, 62–63, 95; lips, 200; males, 95, 179; mentoplasty, 62; otoplasty, 62; rhinoplasty, 62, 194–95; rhytidectomy (face lifts), 62
Crime, 34, 44, 48
Criminals, 34, 48, 169
Cross-cultural norms: averageness, 7; BMI, 12–13; breasts, 215; cosmetic surgery, 63–64; cosmetics, 200; dental care, 202; differences, 4, 12–13, 27, 150; hair, 166–68; media, 28, 150–51; proportions, 89; relationships, 101; similarities, 3, 36, 115, 128, 178, 208, 215; symmetry, 5, 27; WHR, 208; youth, 53
Cunningham, Michael, 5, 42, 67
Curiousness, 19

Dating, 18, 41, 72–73, 75, 122, 126–27, 132, 168, 189, 192, 214
Death, 32–33, 76, 77, 201
Dental care, 67, 201
Dental health, 201
Deodorant, 68–69, 78, 228–29
Depression, 23, 32, 70, 141–42, 147–49, 154, 156, 159–64, 218, 229
Developmental environment, 5, 22, 27, 32, 42, 176, 192
Developmental stability, 7, 29, 32, 50–51, 167, 232
Diabetes, 32–33, 180, 207, 209
Diet: dental health, 201; dieting, 81–84, 157; dieting in childhood, 142; eating disorders, 159; extreme dieting, 88, 96, 149; healthy diet, 54, 72, 79, 204; inguinal crease, 218; Photoshop, 149; self-esteem, 158, 163; SHR, 210
Divorce, 101, 117, 119–20, 129–33, 139
Dixson, Barnaby, 15, 170–72, 216, 219
Dominance: attraction, 10, 19, 35, 110, 185; baldness, 66, 172–73; beards, 93; BMI, 208; culture, 154;

eyebrows, 175; facial expression, 55, 201–2; facial features, 7, 188, 193, 195; flirtation, 133; height, 212–14; masculinity, 28, 156, 185, 225; posture, 55; scars, 178–79; SHR, 15, 43, 210; symmetry, 66; tattoos, 75; vocal quality, 230

Double eyelid surgery, 62, 154, 190–91

Dove campaign, 83, 89

Ears, 62, 195–97

Eating disorders, 13, 82, 85, 88, 95, 141–42, 146–49, 154, 156–57, 159–61, 163–64; anorexia, 13, 82, 159–60; bulimia, 13, 159

Economic resources, 103–4

Elongated neck, 212

Elongated skulls, 186

Empathy, 18, 21

Employment, 35, 47

Engaged, 126

Environmental circumstances, 97, 106, 111, 113, 116

Environmental stress, 22

Estrogen, 9, 15, 166, 168, 178, 184, 200, 210, 225, 230

Evolutionary perspective, 26, 105, 107, 113, 171, 181, 189

Evolutionary pressures: and attraction, 97, 190; and beauty trends, 81, 88; and hair, 93–94; and reproduction, 100, 115, 190; sexual behavior, 127

Evolutionary psychology, 3, 230

External locus of control, 126

Extramarital affairs, 59, 108, 121, 128

Extraversion, 18–23, 35, 40, 124

Eye: attention, 46, 172; attractiveness, 5, 60, 104; color, 189–90; contact, 118, 192, 223; cosmetic surgery (blepharoplasty), 62–63, 150, 154, 190–92; cosmetics, 56–58, 60–61, 76–78; flirting, 133; gaze, 192; health, 54, 88, 104; iris, 193–94; limbal ring, 193–94; placement, 190–92; sclera, 194; shape; size, 9–11, 42–43, 55, 67, 184, 188, 192–94; spacing, 5; symmetry, 192

Eyebrow: attractiveness, 174–76, 184, 222; flirting, 133; grooming, 56, 61, 67, 87, 173, 175; height, 55, 184; placement, 175; sexual dimorphism, 174–76; signals, 66; thickness, 61

Eyelid surgery, 62, 154, 190–91

Face shape, 5, 8, 183–85

Facial expression, 54–55, 60, 79, 171, 175

Facial hair, 9, 66, 93, 166, 170–71, 184, 193

Facial symmetry: asymmetrical, 192; attractiveness ratings, 5–7, 22–23, 32, 104–5; beards, 66–67, 170, 173; cosmetics, 56; ears, 197; eye symmetry, 192; health, 7, 10; nose, 194; personality, 23; scarification, 177; tattoos, 182–83; testosterone, 203; vocal quality, 230

Faithful, 44, 58, 73, 110, 126, 128–31, 139, 203, 215

Familiarity, 4, 21, 24–26, 106, 188, 228, 231

Fashion models, 28, 83, 146

Fat, 14–15, 32, 88, 91–92, 95, 104, 142, 150, 157, 200, 204–9, 218

Feedback loop, 37, 42, 51, 79, 118, 126, 162–63

Female attractiveness: age, 11; BMI, 12–13, 204; features, 5; hair, 166; health, 49; hormones, 225; nose, 195; SHR, 215; smiling, 203; social ranking, 40; video games, 148. *See* Attractiveness

Female preferences, 9, 103, 169, 212, 218, 222

Femininity: athleticism, 157; attractiveness, 54, 184–85, 225; body, 83; clothing, 47, 74, 85, 157; cosmetic surgery, 63; cosmetics, 61–62, 84, 155, 200; cultural differences, 150–51; facial features,

188, 193, 200; feet, 219–21; fertility, 185, 225; flirtation, 133; hair, 67, 166–69, 173–75; hands, 70, 204; health, 86; in males, 8–10, 28, 43, 179, 186; media, 145, 150–51; relationship satisfaction, 125; sexual dimorphism, 5, 8; tattoos, 75; WHR, 11, 13–14, 32, 106, 208; workforce, 47

Fertile phase, 16, 28, 66, 110, 175, 185, 188–89, 194, 221–27, 230

Fertility: age, 86–87, 104–5, 114; attractiveness, 4, 11, 31, 33, 42, 53, 111, 128, 186, 194, 203; BMI, 205, 208; body scent, 16–17, 223; breasts, 215; complexion, 176, 178, 185; hair, 65–66, 165–68, 170–71, 175; health, 32, 189; menstrual cycle, 27–28; MHC, 226–27; ovulation, 221–25, 230; posture, 56; relationships, 10; sex ratio, 112; sexual interest, 43, 110, 188; skin bleaching, 180; stretched lips, 198; vocal quality, 17; WHR, 14, 32, 209

Financial stability, 87, 103, 130

First impressions, 20, 24

Fitspiration, 159

Flirtation, 117, 133, 139, 224

Foot binding, 220–21

Foot size, 74, 219–21

Foreheads, 46, 188

Fragrances, 57, 69, 228

Friendships, 21, 35, 97, 106–7, 118, 146, 159

Fundamental attribution bias, 48

Gait, 74

Gall bladder disease, 32–33

Gender bias, 47

Gender neutral, 11

Genes, 4–5, 10, 16, 32, 42, 61, 64, 70, 77, 97–98, 103, 107–8, 113–15, 131, 189, 204, 214, 225–27, 230

Genetic quality, 7–8, 22, 26, 28–29, 62, 67, 78, 99, 108, 115, 143, 201

Genetic potential, 22, 50, 67

Genotype, 189

GI Joe, 28

Girl scouts, 83

Golden proportion, 89, 184

Grooming, 61–62, 64–65, 78, 81, 93–94, 172, 175, 178, 201, 204, 224, 232

Gym membership, 71

Hair: age, 67, 88; baldness, 87, 172–73; care, 66, 78, 155, 157; color, 5, 60, 64, 77–79, 168–70; culture, 154–55; length, 5, 64, 91, 166–68; facial, 9, 66, 93, 166, 170–71, 184, 193; health, 10, 54, 64–65, 104, 165; hormones, 94; products, 76, 87, 143, 153, 155; removal, 64–65, 155, 173–74

Hairstyle, 58, 81, 93, 96, 166, 186

Halo effect, 33–35, 38–39, 48–50, 167

Hands, 55, 70–71, 133, 203–4

Handsome, 168–69, 212, 219

Health problems: anemia, 33; cancer, 32–33, 150, 180, 209; diabetes, 32–33, 180, 207, 209; gall bladder disease, 32–33; heart disease, 32–33, 207; thyroid disease, 33; stroke, 32–33, 209

Heart disease, 32–33, 207

Height, 12, 55, 74–75, 83, 89, 93, 188, 204, 211–15

Heterosexual, 42, 161

High energy, 10, 104

High heels, 74, 94, 211

High sex ratio, 112–13

Homosexual, 136, 161

Honesty, 21, 24, 28, 43, 49, 66, 122–23, 173

Hormonal influences on attraction, 110, 221, 224–25, 230, 232

Hormones: birth control, 224; body shape, 215; hair, 64, 167; pheromones, 16; sexual attraction, 110, 224–25; SHR, 32–33, 210;

vocal quality, 230; weight, 207; WHR, 14, 32, 209
Hourglass shape, 9, 11–14, 32, 43, 74, 77, 106, 208–9, 217
Humor, 33, 67, 102, 105
Hunter-gatherer, 7
Hygiene, 70, 78–79, 94; oral, 67, 200–201

Identity, 82, 183, 198
Immune system, 9, 15–16, 33, 67, 115, 178, 181, 210, 225, 227, 230
Independence, 35, 136, 189
Infections, 9, 65–67, 155, 174, 178, 201
Infidelity, 132, 189
Inguinal crease, 218
Initiation of relationships, 117–18
Intelligence: and attraction, 24; and attractiveness, 34–35, 42, 49–50, 58, 102–3, 105, 231; and emotional intelligence, 124; facial features, 185–87; and hair length, 64, 167, 173; and opportunities, 44; ratings of, 38; smile, 201
Internalization of ideals, 144–47, 149–50, 152, 158–61, 163–64, 174
Internal locus of control, 126
International Mate Selection Project, 101, 104
Interoceptive awareness, 156
Interpersonal interest, 21
Intimacy: attractiveness, 21; children, 126; emotional, 124, 126–28, 134; femininity, 125; intellectual, 124; marriage, 126; personality, 125, 231; physical, 124, 127–28; impact of pornography, 137–39; quality, 27; relationship satisfaction, 117, 123–26, 139, 228, 231; sharing experiences, 24; sexual, 107, 127; spiritual, 124; Triarchic theory of love, 123; trust, 125
Introverts, 18
Investment: attractiveness, 45; in children, 38, 114, 189; parental investment, 9–10, 28, 38, 98–100, 116, 134, 167; in a relationship, 42, 47, 109–12, 126; relationship satisfaction, 123; reproduction, 108
Iris, 193–94

Japan, 7, 13, 63, 77, 101, 150–51, 180, 190
Jealousy, 35, 43, 49, 117–18, 120–24, 133, 139, 210, 223

Karo tribe, 177
Kayan tribe, 212–13
Kindness, 18, 43, 102, 105, 123
Kissing, 129, 197, 229
Korea, 63, 143, 190

Labiaplasty, 63
Langlois, Judith H., 3, 36, 41, 49
Large bodies in Mauritania, 206–7
Laser hair removal, 64–65
Legs: fashion, 85–86; hair, 65–66, 155, 173–74; length, 83, 104, 210–11; males, 91
Leg-to-trunk ratio (LTR), 210–11
Length of relationship, 27, 116, 231
Limbal ring, 193–94
Lincoln, Abraham, 93
Lipombo, 186
Liposuction, 62–63, 95
Lips: age, 10–11, 42, 193; attraction, 104, 200–201; Botox, 64; color, 56, 77, 200; cosmetic surgery, 200; impact of hormones, 9; lipstick, 57, 61, 77, 203; smile, 55; stretched lips, 197–99
Locus of control, 124, 126; external locus of control, 126; internal locus of control, 126
Loneliness, 21
Long-term relationships, 9–10, 28–29, 34, 44, 69, 75, 87, 97–104, 106–15, 118–19, 123, 128–29, 139, 171, 178, 185, 203, 215–16, 223, 228, 230–32
Low commitment, 121–22, 216

Low sex ratio, 112–13
Lumbersexual, 93

Maasai, 195–97
Macaque, 5, 223
Major histocompatibility complex (MHC), 225–28
Male attractiveness: age, 12; BMI, 12–14; body hair, 66; features, 5; health, 49; hormones, 224–25; muscles, 92; SHR, 210; WHR, 209. *See* Attractiveness
Male genitalia, 218
Male preferences, 66, 169, 205, 230
Malnutrition, 32, 105
Mangbetu tribe, 186–87
Marriage, 26–27, 78, 100–102, 119–21, 130–32, 138, 167, 182–83, 196–99, 205–6, 221
Masculinity: attractiveness, 7–8, 19, 28, 54, 108, 185, 200, 222; baldness, 172–73; cosmetics, 180; culture, 81; facial features, 188, 200, 225; facial hair, 66–67, 93, 166, 172, 175, 184–85; fathers, 110; flirtation, 133; hair, 165–66, 171; intimacy, 125; media, 91–92, 95, 148; muscles, 218; personality, 43; relationships, 110–11; scars, 178; sexual dimorphism, 5, 8–10, 184; SHR, 15–16, 90, 210; tattoos, 75–76; testosterone, 171; vocal quality, 230; WHR, 14; in women, 47, 65, 173
Mate poaching, 117–19, 132, 139
Mating strategy, 215
Mating style: monogamy, 108, 130; polyandry, 111; polygamy, 101; polygyny, 99, 111–12
Mauritania, 206–7
Media: age, 176; beauty ideals, 28, 81, 97, 141, 174; BMI, 12, 205; cosmetic surgery, 88; cultural differences, 143, 152–55; dental ads, 202; diversity, 153; Dove campaign, 83; facial hair, 93; fashion magazines, 145–46; Girl Scouts, 83; muscularity, 92, 95, 217; music videos, 149; Photoshop, 88, 149–50; psychological effects, 141–42, 156–64; sex appeal, 95; television and movies, 146–48; video games, 148; WHR, 208
Menopause, 14, 32, 105, 209
Menstrual cycle: attraction, 8–10, 171, 194, 225; attractiveness, 185, 222; health, 32; scent, 16, 227
Mental health, 5, 8, 13, 41, 51, 96, 195, 160, 162, 197
Mentoplasty, 62
Mere exposure effect, 26
Metrosexual, 70, 93–94, 179
Michelangelo's *David*, 90–91
Miss America, 13, 28, 83
Modesty, 85, 150
Monogamy, 101, 108, 111–14, 129–30, 215
Monroe, Marilyn, 13
Motivation: attractiveness, 21, 50, 141, 222; career, 72; divorce, 133; gender, 107; to internalize ideals, 164; relationships, 114; scent, 229; shaving, 174
Mursi tribe, 198–99
Muscularity: age, 10–11, 87; androgens, 208; attractiveness, 9, 72, 90; body anxiety, 147, 159; fashion, 73, 211; female preferences, 10, 13, 28, 72, 91, 217–18; GI Joe, 28; gym, 71; health, 88–89, 217–18; male preferences, 104, 145; media, 148, 159–61; sex differences, 54, 81, 85, 217–18; SHR, 15; trends, 91–96

Nail care, 70, 157, 203–4
Natural beauty, 54–56, 76, 79
Natural selection, 98, 116
Neuroticism, 18–23, 40, 124, 201
Nonindustrialized societies, 12, 205
North America, 7, 14, 122, 170

Nose: attractiveness, 46, 192, 227; cosmetic surgery (rhinoplasty), 62–63; difference in ideal between genders, 5, 9, 11, 184, 194–95; scent, 227, 229; size, 43
Nutrition, 5, 22, 32–33, 42, 105, 166, 211
Nylons, 83, 86

Obesity, 27, 32–33, 105, 146, 205, 207
Objectification, 145, 147, 149–50, 156
Objective qualities, 3, 6, 25, 32, 132, 232
Openness to new experience, 18–21
Opportunities: educational, 39; future, 31, 70, 79, 141–42, 162, 208; occupational, 34–35, 41–42, 44–46, 48, 50, 157, 214; relationships, 119, 192; reproductive, 32–33, 55, 77, 112, 139; social, 28, 33, 40, 51, 86, 115, 117
Oral hygiene, 67, 200–201
Otoplasty, 62
Ovulation: attraction, 221–24; scent, 16, 223; sexual interest, 110; vocal quality, 230; WHR, 32

Pageants, 88
Papua New Guinea, 177, 182–83
Parasites, 178,
Parental investment, 9, 28, 38, 98–100, 116, 134, 167
Penis, 219
Performance reviews, 48
Personality: attraction, 4–5, 17–18, 20–24, 27–29, 67, 165–67, 201, 231; big-five, 20–22, 40; halo effect, 35, 118; kissing, 229; relationship length, 231; relationship satisfaction, 117, 123–24; relationship type, 108, 114; similarities, 24, 40; stability, 8; symmetry, 22–23; tattoos, 75; women, 82
Phenotype, 189

Pheromones, 16, 65, 69, 222, 228–29. *See also* Body scent
Photoshop, 88, 115, 149–50, 163, 205
Physiology, 210
Playboy, 13, 28, 169
Politics, 34, 101
Polyandry, 111
Polygamy, 101, 104
Polygyny, 99, 111–12
Popularity, 36–40, 49–50, 72, 167
Pornography, 66, 117, 134–39, 148
Positive reinforcement, 142
Posture, 54–56, 61, 74, 79, 83
Pregnancy, 32, 100, 105, 107, 109, 137, 209, 215, 227
Prisoner's dilemma, 44–45
Promiscuity, 75–76, 107–8, 111–13, 115, 203, 216
Propinquity, 24–26
Proportions: body, 3, 86, 96, 204–13; facial, 6, 11–12, 188; golden, 89; ideal, 91, 95; muscles, 28; Photoshop, 150
Prosocial, 19
Psychological characteristics, 5, 108. *See* Personality
Psychological health, 23, 51, 85, 95, 136, 146, 161–62, 204, 218
Puberty, 14, 17, 178, 184, 209–10, 215
Punishments, 44, 48

Qualifications, 39, 46–47
Quality: attractiveness, 110; body scent, 28, 70; food, 106; genetic, 7–8, 22, 26, 28–29, 62, 67, 78, 99, 108, 115, 143, 201; hair, 64, 165–67, 232; hormones, 110, 184; intimacy, 27; of life, 77, 79; offspring, 113, 116; partners, 44, 73, 100, 108, 177; personality, 34–35; relationship, 25, 103, 117–26, 130, 137, 139; sex, 128; skin, 56, 165, 181–83, 204, 232; social interactions, 51; sperm, 8, 33; teeth, 67, 200; vocal, 5, 8, 17, 230–31

Race: and attraction, 24, 101, 181; and cosmetic surgery, 154; ideals, 165; and media, 157; in the modeling industry, 64, 149, 151–52; of the perceiver, 3, 7, 37, 46, 142

Relationships: adulthood, 41; asymmetry, 23; childhood, 36; commitment, 27, 113–14, 127–29, 171; cultural differences, 112–13, 116; divorce, 129–33; eating disorders, 150–62; empathy, 18; familiarity, 25; fantasies, 133–35; flirtation, 133; friendship, 106; gender differences, 100–101, 104, 107–11, 114–15, 185, 189, 216, 223; hormones, 203, 223, 226–28, 230; initiation, 117–18; intimacy, 27, 55, 123–27, 174; jealousy, 120–23; length, 10, 27, 69, 87, 97, 100, 107–11, 112, 118–20, 231–32; maintenance, 35, 117, 222; monogamy, 112, 215; opportunities, 44, 72, 86; personality, 29; pheromones, 16, 229; pornography, 135–39; quality, 24, 49, 170; romantic, 42, 49, 97; satisfaction, 69, 117, 123, 127; sexual experience, 19, 75, 100; stability, 19, 69; type, 21, 97, 101, 106, 111–12, 185

Religion, 24, 74, 78, 101, 130

Renzao meinü, 63

Reproduction: age, 10–11, 64, 76, 87, 98–99, 105, 178, 197, 203; attraction, 4, 106; BMI, 12, 33, 50, 88–89, 204, 208–9; breasts, 216; commitment, 127; cosmetic use, 76–77; cultural differences, 4; fertility, 66, 87, 105, 166; fitness, 10, 96; gender differences, 103, 115, 134; health, 50, 99, 189, 216, 232; height, 214–15; hormones, 33, 110, 185, 188, 227, 230; muscularity, 99; offspring, 16; parental investment, 99–100, 108–9, 115, 189; personality, 18; potential, 31–32; scent, 17, 223–24, 228–29, 232; sex ratio, 112; sexual selection, 98–100; short-term mating, 108; similarities, 26, 107; social exchange theory, 113, 115; status, 82; success, 33, 61, 88; WHR, 14, 33, 50, 106, 208–9

Resources: attractiveness, 72, 128, 131; commitment, 123; cultural differences, 4; cultural similarities, 102–3; displaying resources, 72; divorce, 131; and extramarital affairs/infidelity, 108–9, 132; male investment, 10, 73, 99, 102–4, 107, 109, 111, 189; metabolic, 9; monogamy, 114; polygamy, 112; sex ratio, 112; skin color, 180; social exchange theory, 113–14; society, 4, 12, 91

Respect, 21, 37, 40, 49, 50, 55, 75, 123, 125, 127, 195–99

Rhinoplasty, 62, 194–95

Rhytidectomy, 62

Risk taking behavior, 42, 75, 91, 99, 110, 115, 178, 223

Ritual scarring, 177

Rude, 21

Salary, 47

Scars, 61, 177–78, 183

Sclera, 193–94

Self-assessment, 27, 42

Self-consciousness, 68, 82, 147, 160, 164, 201

Self-efficacy, 38, 51, 126

Self-esteem: age, 54, 87; anxiety, 163; attractiveness, 32, 34, 42, 51, 115, 117, 125; beauty ideals, 88–89, 95, 157–59; BMI, 207–8; children, 38; college students, 39; cosmetic surgery, 63–64; cultural differences, 152–54; eating disorders, 159; fat, 218; intimacy, 124; jealousy, 120–21; mate poaching, 132; media, 141, 146, 148–49, 152; muscularity, 218; pornography, 137; positive

expectations, 37; relationship satisfaction, 125; respect, 50; smiling and teeth, 68, 201–3; viewing attractive others, 27
Self-fulfilling prophecy, 28, 38. *See also* Feedback loop
Self-objectification, 145, 147, 149
Sense of humor, 33, 67, 102, 105
Sex ratio, 97, 112, 116
Sexual dimorphism, 8–10; attractiveness, 27; cosmetics, 61, 65; facial features, 184; fashion, 74; hair, 166, 170; vocal quality, 17; WHR, 14
Sexual disorders, 156
Sexual experience, 19, 41, 105, 112, 114–15, 135–38
Sexual fantasies, 117, 133–35, 139
Sexual satisfaction, 126–27, 135–36, 138–39
Sexual selection, 98–102, 116
Shared values, 102, 125
Short-term relationships, 10, 97, 101, 106–10, 113, 115, 119, 128–29, 171, 178, 185, 203, 216, 230
Shoulder pads, 73, 83
Shoulder-to-hip ratio (SHR), 13–15, 17, 19, 33, 43, 54, 89–92, 95, 210, 215, 230
Similarity: attractiveness, 24–28; cross cultural, 5, 53, 165; genetic, 106–7, 226; internalization of ideals, 149; MHC, 226–27; sexual dimorphism, 9; subjective qualities, 4, 231
Singh, Devandra, 11, 13
Single eyelid, 190
Sister Wives, 101
Skin: age, 53, 56, 87, 166; attractiveness, 104, 176–83, 223, 230; bleaching, 143–44, 156, 180; color, 53, 64, 81–82, 92, 153–56, 179–81; complexion, 176, 178–79; cosmetic surgery, 62; cosmetics, 56, 61, 69, 76–77, 93; cultural differences, 53; eyelids, 190–92; fashion, 74, 85, 224; hair removal, 65; hands, 70, 203–4; health, 10, 54, 72; hygiene, 91; media, 145; scarring, 177; tattoos, 75, 182–83; texture, 5, 8–9, 88, 173, 181–83; tone, 95–96, 222
Sleep deprivation, 54
Smile, 5, 9, 42–43, 55, 67–68, 201–2
Social appearance anxiety, 163–64
Social exchange theory, 113–16
Social psychology, 4, 24, 28, 113
Social skills, 18, 23, 39, 41–42, 51, 60, 118, 168
Social status: attractiveness, 103; beards, 172; body weight, 27, 82, 94, 208; cosmetics, 62–63; depression, 162; elongated skulls, 186–87; fashion, 93–94; foot size, 221; hands, 70–71; infidelity, 127–29; jealousy, 120–21; men, 89; muscularity, 91–95; peers, 38, 40; personality, 40; scars, 177; SHR, 92; skin color, 180; stretched earlobes, 195–97; stretched lips, 198; symmetry, 23; tattoos, 75, 78, 183; uniforms, 74; women, 73, 103–6, 114. *See also* Socioeconomic status
Societal pressures, 83, 86, 89, 176
Socioeconomic status (SES), 5, 33, 73, 82, 104, 128–29, 137, 172, 179, 214
Sociosexual orientation, 112–13, 116, 215–16
STDs, 65–66
Stereotypes, 20, 47, 118, 158
Strategic pluralism theory, 112–13, 116
Stretched earlobes, 195–97
Stretched lips, 198
Stroke, 32–33, 209
Submissive, 43, 175, 193, 201
Swami, Viren, 170, 189, 216
Swimwear, 85
Symmetry. *See* Body symmetry; Facial symmetry

Tapeworms, 77
Tattoos, 75–76, 78, 182–83
Teacher evaluations, 35
Teeth, 67–68, 78, 200–203
Testosterone, 9, 15, 20, 66, 94, 98, 110, 134–35, 184–86, 203, 210, 222, 225
Thinspiration, 159
Threat, 19, 27, 120–22, 132–33, 157, 175, 185, 223, 225
Thyroid problems, 33
Toxins, 22
Traditional sex roles, 27
Triarchic theory of love, 123
Trustworthiness, 23, 35, 45–46, 89, 107, 126, 185, 225
T-shirt test, 17, 227, 230
Twiggy, 13
Type II diabetes, 32–33
Type of relationship, 10, 97, 101, 106, 129, 139

Underweight: attractiveness, 205–6; BMI, 204–5; cultural pressures, 13; health, 27, 33, 159, 228; models, 88, 145, 205; poverty, 91
Uniforms, 73–74
United Kingdom, 63, 171, 216
United States: age, 88; beauty trends, 4, 56, 91; birth control, 107; BMI, 12, 106; body image, 156–57, 204, 207; cosmetic surgery, 62–64, 195, 200; cosmetic use, 53–54; dental care, 67–68; depression, 142; divorce, 130; eating disorders, 159; fashion, 73–74; gym membership, 71; hair length, 166–68; hair removal, 65–66, 94; hand care, 70; marriage, 101, 130; media, 81, 142–43; models, 25, 150–54; perceptions, 34–35, 141; pornography, 135–36; skin, 179; preferences, 11–12, 53, 210, 219
Universal characteristics of beauty, 4

Ventral occipital region of visual cortex, 27, 118

Victorian era, 86, 221
Vitruvian man, 89
Vocal quality, 3, 5, 17, 222, 230

Waist-to-chest ratio (WCR). See Shoulder-to-hip ratio (SHR)
Waist-to-hip ratio (WHR): affairs, 119; age, 105; cross cultural consistency, 208–10; female attractiveness, 11, 21, 43, 50, 104–6, 207, 215, 217; health, 14, 32; hormones, 32–33; ideal, 13–15, 81, 89; males, 14–15, 92, 210; reproduction, 209; vocal quality, 17
Waist-to-shoulder ratio (WSR). See Shoulder-to-hip ratio (SHR)
Weight: age, 11, 87; attractiveness, 54; BMI, 12, 204, 206, 208; bullying, 218; cultural differences, 152, 159; dieting, 142; eating disorders, 159; fertility, 32–33, 105; Girl Scouts, 83; health, 12, 27, 33, 88; ideal weight, 81–82, 145, 163; marriage, 132; models, 13, 56, 88, 145; race, 13; reproduction, 228; status, 91, 95, 106; weight loss, 142, 205
Well-heeled, 94
Wingman, 72
Workplace, 34–35, 41, 48
Wrinkles, 10, 54, 56, 61, 64, 87, 178

Youth, 10–11; and attraction, 5, 7, 18, 42–43, 47, 76, 88, 91, 104, 231; beauty, 87; chin, 188; complexion, 57, 77; and cosmetic surgery, 64, 87; and cosmetics, 56, 61–62; eyebrows, 175; eyes, 193; fertility, 31–32, 114, 178; foot size, 221; and hair, 64–67, 166, 168, 173; lips, 200; models, 151; muscularity, 217; nose, 195; preservation of, 53, 56; and reproduction, 99, 112; skin, 166, 181; smile, 201; tattoos, 78; teeth, 202; WHR, 209

About the Author

Rachelle M. Smith, PhD, earned her doctoral degree in psychology at the University of Maine, Orono, in 2009. She is an associate professor of psychology and the chair of social sciences at Husson University. Dr. Smith teaches a variety of psychology courses and specializes in evolution and development. Her published works include journal articles and book chapters in the areas of mate choice and reproductive strategies, tactical deception, and social intelligence. Dr. Smith received the Owen Aldis Research Award and is a member of the International Society for Human Ethology.